algebraforall red level

elizabeth warren PhD

Series Consultants
James Burnett MEd
Calvin Irons PhD

About the Author

Elizabeth Warren has been involved in Mathematics Education for more than 30 years. During this time she has actively worked in both the tertiary and school levels and engaged both elementary and secondary schoolteachers in professional development activities. She is presently conducting research into Patterns and Algebra.

Algebra for All, Red Level

Copyright 2007 ORIGO Education
Author: Elizabeth Warren PhD
Series Consultants: James Burnett MEd and Calvin Irons PhD

Warren, Elizabeth.
Algebra for All: red level.

ISBN 1 921023 05 8.
1. Algebra - Problems, exercises, etc. - Juvenile literature. I. Title.
512

For more information, contact
North America
Tel. 1-888-ORIGO-01 or 1-888-674-4601
Fax 1-888-674-4604
sales@origomath.com
www.origomath.com

UK/Australasia
info@origo.com.au
www.origo.com.au

All rights reserved. Unless specifically stated, no part of this publication may be reproduced, or copied into, or stored in a retrieval system, or transmitted in any form or by any means, electronic, mechanical, photocopying, recording, or otherwise, without the prior written permission of ORIGO Education. Permission is hereby given to reproduce the blackline masters in this publication in complete pages, with the copyright notice intact, for purposes of classroom use authorized by ORIGO Education. No resale of this material is permitted.

ISBN: 978 1 921023 05 7

10 9 8 7 6 5 4 3 2 1

INTRODUCTION 2

Equivalence and Equations
Everyday Equations 6
Pocket Money 8
More or Less 10
Dollar Dazzlers 12
Keeping It Balanced 14
Making Connections 16
Mystery Masses 18
Boxed In 20

Patterns and Functions
Just Juice 22
Juicy Juice 24
Number Rectangles 26
Frieze Frame 28
Bulk Buys 30
How Much? How Many? 32
What's the Rule? 34
Forward, Backward 36
Odds or Evens 38

Properties
Partial Products 40
Look Both Ways 42
Rearranging Rectangles 44
No Change 46
Following Orders 48
Building Blocks 50

Representations
High in the Sky 52
Card Games 54
Prize Money 56
Fill It Up 58
It Depends 60
The Great Escape 62
Scale It 64

ANSWERS 66

INTRODUCTION

What is algebra?

Algebraic thinking commences as soon as students identify consistent change and begin to make generalizations. Their first generalizations relate to real-world experiences. For example, a child may notice a relationship between her age and the age of her older brother. In the example below, Ali has noted that her brother Brent is always 2 years older than her.

Ali's age	Brent's age
8	10
9	11
10	12
11	13

Over time these generalizations extend to more abstract situations involving symbolic notation that includes numbers. The above relationship can be generalized using the following symbolic notation.

Ali + 2 = Brent   A + 2 = B

Algebraic thinking uses different symbolic representations, such as unknowns and variables, with numbers to explore, model, and solve problems that relate to change and describe generalizations. The symbol system used to describe generalizations is formally known as algebra. Following the example above, Ali wonders how old she will be when Brent is 21 years old. We can solve a problem such as this by "backtracking" the generalization (A = 21 – 2) or using the balance method of subtracting 2 from both sides the equation (Ali = Brent – 2).

Why algebra?

Identifying patterns and making generalizations are fundamental to all mathematics, so it is essential that students engage in activities involving algebra. Many practical uses for algebra lie hidden under the surface of an increasingly electronic world — specific rules are used to determine telephone charges, track bank accounts and generate statements, describe data represented in graphs, and encrypt messages to make the Internet secure. Algebraic thinking is more overt when we create rules for spreadsheets or simply use addition to solve a subtraction problem.

> Algebra involves the generalizations that are made regarding the relationships between variables in the symbol system of mathematics.

What are the "big ideas"?

The lessons in the *Algebra for All* series aim to develop the "big ideas" of early algebra while supporting thinking, reasoning, and working mathematically. These ideas of equivalence and equations, patterns and functions, properties, and representations are inherent in all modern curricula and are summarized in the following paragraphs.

Equivalence and Equations

The most important ideas about equivalence and equations that students need to understand are:

- "Equals" indicates equivalent sets rather than a place to write an answer
- Simple real-world problems with unknowns can be represented as equations
- Equations remain true if a consistent change occurs to each side (the balance strategy)
- Unknowns can be found by using the balance strategy.

Patterns and Functions

This idea focuses on mathematics as "change". Change occurs when one or more operation is used. For example, the price of an item bought on the Internet changes when a freight charge is added. It is important for students to understand that:

- Operations almost always change an original number to a new number
- Simple real-world problems with variables can be represented as "change situations"
- "Backtracking" reverses a change and can be used to solve unknowns.

Properties

Students will discover a variety of arithmetic properties as they explore number, such as:

- The commutative law and the associative law exist for addition and multiplication but not for subtraction and division
- Addition and subtraction are inverse operations, as are multiplication and division
- Adding or subtracting zero and multiplying or dividing by 1 leaves the original number unchanged
- In certain circumstances, multiplication and division distribute over addition and subtraction.

Representations

Different representations deepen our understanding of real-world problems and help us identify trends and find solutions. This idea focuses on creating and interpreting a variety of representations to solve real-world problems. The main representations that are developed in this series include graphs, tables of values, drawings, equations, and everyday language.

INTRODUCTION

About the series

Each of the six *Algebra for All* books features 4 chapters that focus separately on the "big ideas" of early algebra — Equivalence and Equations, Patterns and Functions, Properties, and Representations. Each chapter provides a carefully structured sequence of lessons. This sequence extends across the series so that students have the opportunity to develop their understanding of algebra over a number of years.

About the lessons

Each lesson is described over 2 pages. The left-hand page describes the lesson itself, including the aim of the lesson, materials that are required, clear step-by-step instructions, and a reflection. These notes also provide specific questions that teachers can ask students, and subsequent examples of student responses. The right-hand page supplies a reproducible blackline master to accompany the lesson. The answers for all blackline masters can be found on pages 66-73.

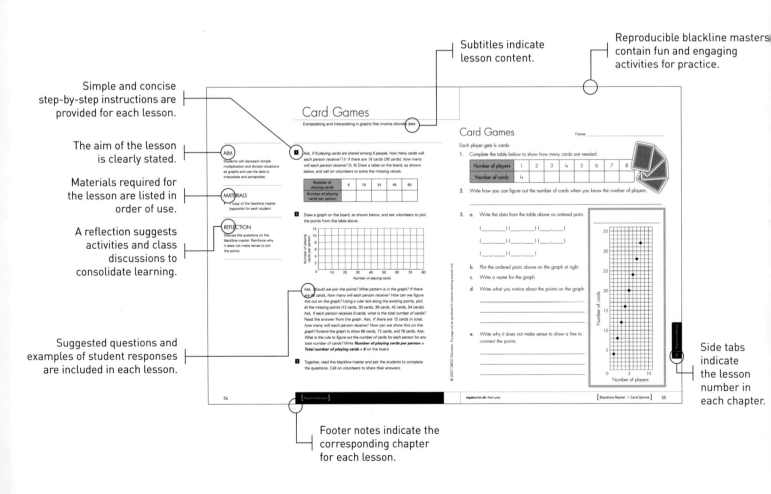

Assessment

Students' thinking is often best gauged by the conversations that occur during classroom discussions. Listen to your students and make notes about their thinking. You may decide to use the rubric below to assess students' mathematical proficiency in the tasks for each lesson. Study the criteria, then assess and record each student's understanding on a copy of the Assessment Summary provided on page 74. Although the summary lists every lesson in this book, it is not necessary to assess students for all lessons.

A	The student fully accomplishes the purpose of the task. Full understanding of the central mathematical ideas is demonstrated. The student is able to communicate his/her thinking and reasoning.
B	The student substantially accomplishes the purpose of the task. An essential understanding of the central mathematical ideas is demonstrated. The student is generally able to communicate his/her thinking and reasoning.
C	The student partially accomplishes the purpose of the task. A partial or limited understanding of the central mathematical ideas is demonstrated and/or the student is unable to communicate his/her thinking and reasoning.
D	The student is not able to accomplish the purpose of the task. Little or no understanding of the central mathematical ideas is demonstrated and/or the student's communication of his/her thinking and reasoning is vague or incomplete.

Everyday Equations

Matching equations and real-world situations that involve multiplication

AIM

Students will write simple multiplication stories to match equations with one unknown factor, and vice versa. They will also informally relate multiplication and division.

MATERIALS

- 1 copy of the blackline master (opposite) for each student

REFLECTION

Ask, *What words can we use to write word problems for multiplication with unknowns? How can we figure out the missing number?* Encourage the students to suggest contexts that use language such as "four times". For example, "Anne has $36. Anne has four times the amount that Ben has. How much money does Ben have?"

1 Draw the picture shown below on the board.

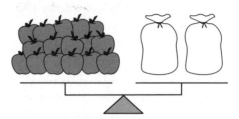

Say, *The number of apples in each bag is the same. What do you know about the apples and bags?* (They balance.) *What equation can we write?* Encourage volunteers to suggest equations that use addition (16 = ☐ + ☐) or multiplication (16 = 2 × ☐) and write them on the board. Ask, *What is the missing number? How do you know?*

2 Write **4 × ☐ = 20** on the board. Ask, *What stories can we write to match this equation?* Encourage the students to suggest real-world stories using a range of contexts (people in buses, birds in trees, boxes of fruit) and draw pictures to match. For example, "Twenty people are being transported in 4 cars with the same number of people in each car. How many people are in each car?"

3 Have the students complete the blackline master. Ask volunteers to share their answers.

[Equivalence and Equations]

Everyday Equations

1. Liam has 3 bowls with the same number of apples in each. Madison has 18 apples. Liam and Madison have the same number of apples.

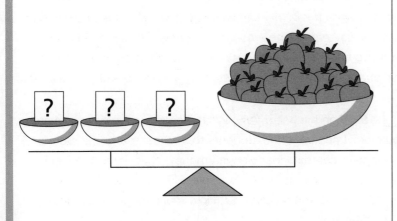

a. Write numbers in this balance picture to match the story.

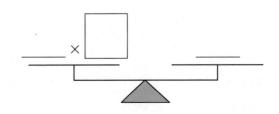

b. Write the equation.

_____ × ☐ = _____

2. a. Write a story for ? × 4 = 36. _____

   _____

   _____

   b. Draw a picture to show your story.

   c. Write numbers in this balance picture to match your story.

   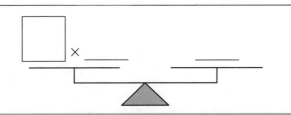

Pocket Money

Using balance situations to write equations that relate multiplication and division

AIM

Students will write simple division stories to match equations with one unknown factor, and vice versa. They will also informally relate division and multiplication.

MATERIALS

- 1 copy of the blackline master (opposite) for each student

REFLECTION

Ask, *How do we know that a story is about division? What words can we use to describe division?* (Share among; split between.) *What equations can we write?* Encourage the students to explain that an equation involving division can be rewritten as a multiplication number sentence.

1 Say, *Ben has $20 in his pocket. Julio has some money in his pocket. If Julio equally shared his money among 4 friends and himself, he will have the same amount of money as Ben.* Discuss questions such as, *How much money does Ben have in his pocket? How much does Julio have in his pocket? Among how many people does Julio share his money? How can we write this? How can we show that, after the sharing, Ben and Julio will have the same amount of money in their pockets?* Invite individuals to draw pictures and write a multiplication and a division equation on the board to match the story. ($\square = 5 \times 20$ and $\square \div 5 = 20$.)

2 Write $\square \div 5 = 7$ on the board. Ask, *What stories can we write to match this equation?* Encourage the students to suggest real-world stories using a range of contexts (people with money, chocolates in boxes) and draw a picture to match. For example, "Five people shared some money and each person received $7. How much money was there before the sharing?"

3 Have the students complete the blackline master. Ask volunteers to share their answers.

[Equivalence and Equations]

Pocket Money

Name _____

1. Ada has $60. If she shares it equally among 4 friends, how much will they each receive?

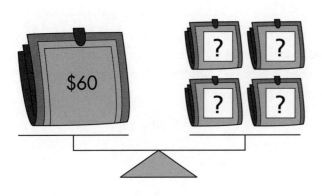

a. Write numbers in this balance picture to match the story.

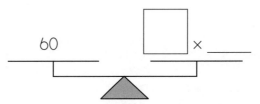

b. Write two equations.

_____ ÷ _____ = ☐

2. a. Write a story for 36 ÷ 3 = ?. _____
   _____
   _____

   b. Draw a picture to show your story.

   c. Write numbers in this balance picture to match your story.

   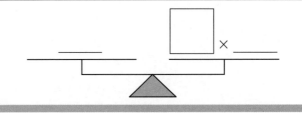

More or Less

Using division to create inequalities

AIM

Students will work with "greater than" and "less than" in situations that involve division.

MATERIALS

- 1 copy of the blackline master (opposite) for each student

REFLECTION

Refer to Question 1 on the blackline master and ask, *How did you figure out the value for each unknown? How can you figure out the greatest value for each? And the least value for each?*

1 Draw the unbalanced scale shown below on the board.

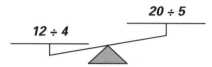

Ask, *Is this true or not true? How do you know? Which side should be greater? How can we make it true by changing only one number?* (For example, change 5 to 10.) Alter the balance picture to reflect the number change, as shown below.

Ask, *What number sentence can we write to show what we see in the picture?* Invite individuals to suggest and write inequalities such as **12 ÷ 4 > 20 ÷ 10** on the board. Then ask, *Is there another way we can write the sentence?* Write **20 ÷ 10 < 12 ÷ 4** on the board.

2 Draw an unbalanced scale on the board with ◯ **÷ 6** on the left side and **42 ÷ 7** on the right. Point to the circle and ask, *What numbers can we write in the circle to make this true? What is the greatest (least) whole number that will make this true?* Write ◯ **÷ 6 ___ 42 ÷ 7** on the board and ask, *What symbol should we write in this number sentence to make it true?* Write **42 ÷ 7 ___ ◯ ÷ 6** on the board and ask, *What symbol should we write in this number sentence to make it true?*

3 Have the students complete the blackline master. Ask volunteers to share their answers.

[Equivalence and Equations]

More or Less

Name _____

1. For each balance picture, choose a number below to make the picture true. Write the number.

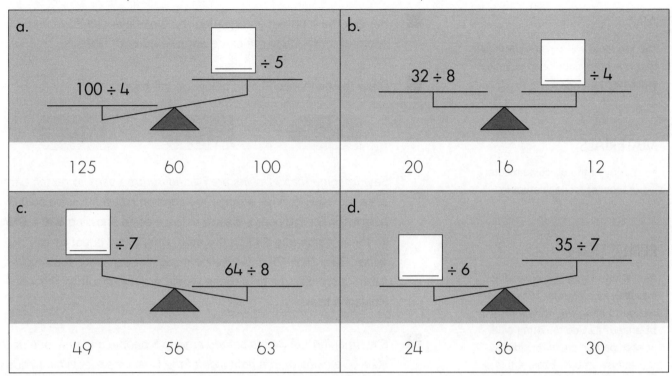

a. 125   60   100

b. 20   16   12

c. 49   56   63

d. 24   36   30

2. Write numbers to make these true. Then write **<** or **>** to complete the matching number sentences.

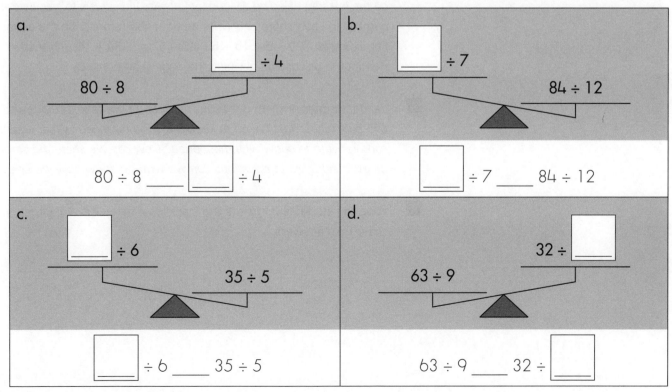

Dollar Dazzlers

Working with equations that involve division

AIM

The students will explore strategies to divide both sides of an equation and keep it balanced.

MATERIALS

- 1 copy of the blackline master (opposite) for each student

REFLECTION

Ask, *If we divide one side of an equation by a number, what must we do to keep the equation balanced?* (Divide the other side, or each part on the other side, by the same number.) Explore this strategy for large positive numbers.

1 Review the fact that an equation remains balanced if we add, subtract, or multiply the same amount on both sides.

2 Draw the price tags shown below on the board.

Say, *Imagine that $1 coins are placed in three bags to match each price tag. How can we arrange the bags on a balance scale so it is balanced?* On the board, draw a balance scale showing **$30 + $15** on the left side and **$45** on the right. Ask, *What equation can we write?* (30 + 15 = 45) If the students suggest equations that involve subtraction, discuss how these would prove difficult to show on a balance scale.

3 Say, *Imagine the $45 is shared among 5 people and only one of the shares remains on the right side of the balance scale. What needs to happen to the amounts in the two bags on the left side to keep the scale balanced?* Encourage students to explain that the two amounts on the left must also be divided by 5. Invite individuals to calculate the number in each share and write an equation to describe the action. For example, "30 ÷ 5 + 15 ÷ 5 = 45 ÷ 5" or "(30 + 15) ÷ 5 = 45 ÷ 5". Repeat the discussion, dividing the original amounts by 3.

4 Lead a discussion with questions such as, *If we take $10 out of the $45 bag, what must we do to keep the scale balanced? What equations can we write to show what happened? How can we share the amounts on each side? What equations can we write to show how we shared?*

5 Have the students complete the blackline master. Ask volunteers to share their answers.

[Equivalence and Equations]

Dollar Dazzlers

Name _____

1. Read the story that matches the picture below.
   4 students each bought a juice and a burger.
   In total, they spent $12 on juices and
   $16 on burgers.

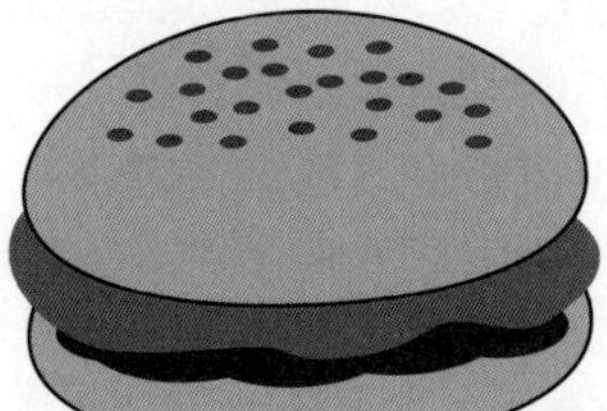

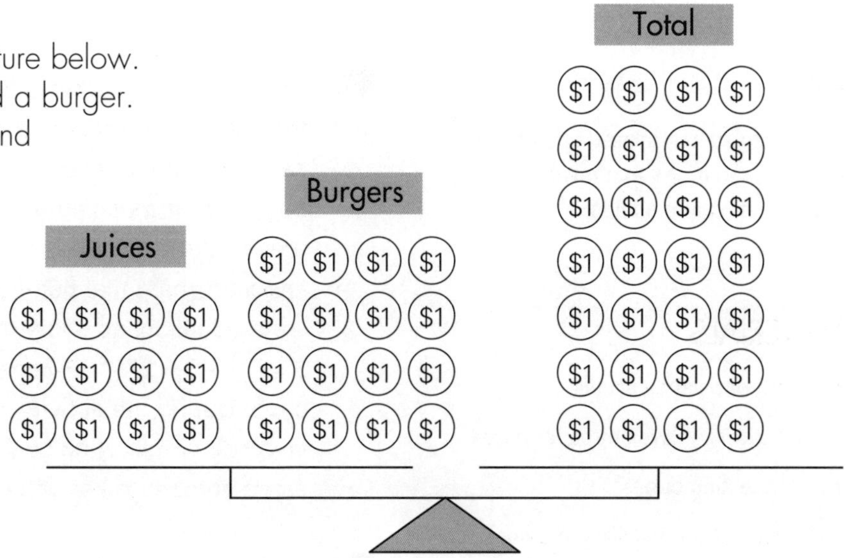

   a. Use the different methods below to figure out each person's share of the total cost.

A Add the amounts spent on each item and divide by 4. Complete the equation.	B Divide the amount spent on juices by 4, then divide the amount spent on burgers by 4. Add both answers. Complete the equation.
(12 + ◯) ÷ 4 = ☐	(12 ÷ 4) + (◯ ÷ 4) = ☐

   b. Write what you noticed. _____
   _____

2. For each balance picture, write two equations to show two methods of figuring out the cost of each person's share. Four people shared the cost.

   a. $20 $28 $48

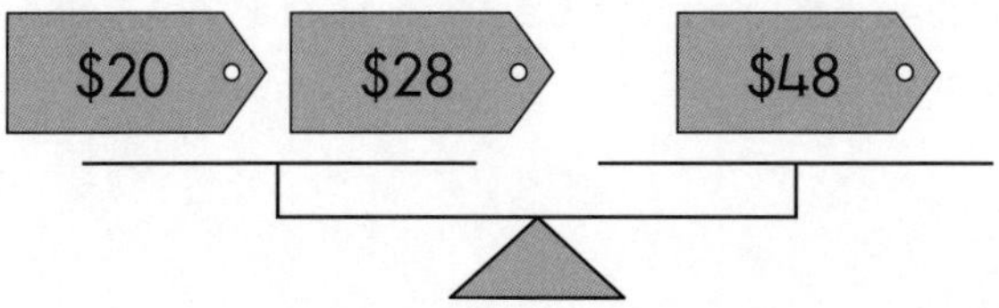

   (20 + ____) ÷ ____ = ____

   b. $32 $40 $72

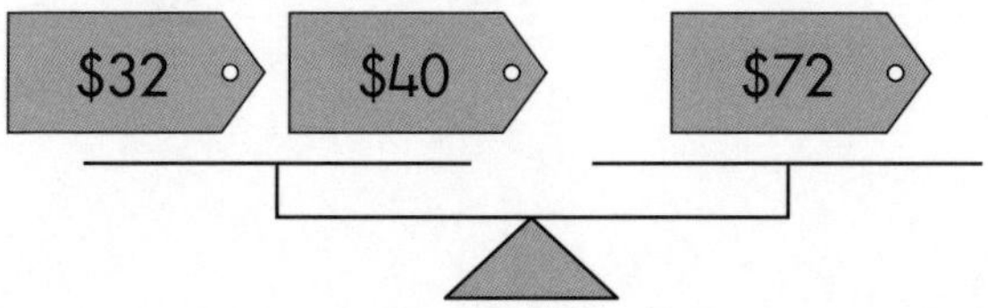

Keeping It Balanced

Using balance to solve problems that involve one or two operations and one unknown

AIM

Students will use related operations to solve equations that involve multiplication and addition.

MATERIALS

- 1 set of balance scales
- Identical lightweight paper bags
- Connecting cubes
- 1 copy of the blackline master (opposite) for each student

REFLECTION

Ask, *How do we use balance to figure out the value of the unknown in multiplication equations?* (Divide.) *How do we check that our answer is correct?* (Substitute the solution in the equation.)

1 Without being observed, place 9 cubes in each of 3 paper bags. Place the bags on one side of the balance scale. Say, *Each bag contains the same number of cubes. How can we figure out the number in each bag without touching the bags?* Encourage students to explain that they could place cubes on the other side of the scale until it balanced and then share that number of cubes by 3. Encourage the students to write equations to show their thinking. (For example, $3 \times \square = 27$ and $3 \times \square \div 3 = 27 \div 3$.) Repeat the discussion for 4 bags with 7 cubes in each. During the activity, reinforce the idea that dividing by 4 is the inverse of multiplying by 4, and that both sides must be divided by the same number to keep the scale balanced.

2 Draw the picture shown below on the board and explain that the number of cubes in each bag is the same. Encourage the students to describe what they know and then write an equation to match on the board. (For example, *$3 \times \square + 2 = 20$*.)

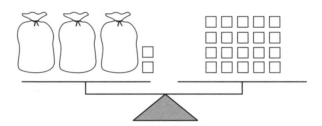

Ask, *How can we change the picture to figure out the unknown number?* (Subtract 2 cubes from each side so there are 3 bags on the left side and 18 cubes on the right. Then divide both sides by 3 so that one bag is on the left and 6 cubes are on the right.)

3 Have the students complete the blackline master. Ask volunteers to share their answers.

Keeping It Balanced

Name _____

1. Complete the steps to find the number of counters in one box.

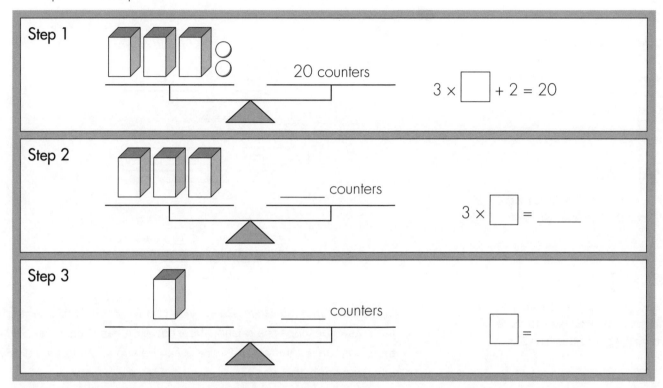

2. Follow the same steps as above to find the missing values.

a.	b.	c.
$3 \times \square + 4 = 25$	$6 \times \square + 3 = 39$	$31 = \square \times 7 + 3$
$3 \times \square = \_\_\_$	$6 \times \square = \_\_\_$	$\_\_\_ = \square \times 7$
$\square = \_\_\_$	$\square = \_\_\_$	$\_\_\_ = \square$
d.	e.	f.
$46 = 4 + \square \times 6$	$5 \times \square - 6 = 29$	$69 = \square \times 8 - 3$
$\_\_\_ = \square \times 6$	$5 \times \square = \_\_\_$	$\_\_\_ = \square \times 8$
$\_\_\_ = \square$	$\square = \_\_\_$	$\_\_\_ = \square$

Making Connections
Identifying the relationship between unknowns

AIM

Students will use logical reasoning to write systems of equations; replace variables with numbers in systems of equations; and identify the relationships between variables.

MATERIALS

- 1 copy of the blackline master (opposite) for each student

REFLECTION

Refer to the blackline master and invite several volunteers to share their solutions to Question 2.

1 Draw the diagrams shown below on the board.

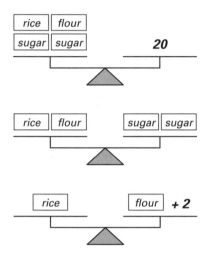

Say, *Look at these balance scales. What do we know?* Encourage the students to describe the relationship shown on each scale. Ask, *What equations can we write to describe what we know?* Write the following equations on the board:

rice + flour + 2 × sugar = 20

rice + flour = 2 × sugar

rice − 2 = flour

2 Ask, *How can we figure out how many units each item weighs?* Refer to the 2nd equation and ask, *One packet of rice plus one packet of flour weighs the same as how many packets of sugar? How many units will 4 packets of sugar weigh?* (20) *How do you know?* (Substitute "2 × sugar" for "rice + flour" in the 1st equation, so 4 × sugar = 20.) Ask, *How much does 1 packet of sugar weigh?* (5 units.) *How can we figure out how much 1 packet of flour weighs?* (Replace the flour in the 2nd equation with "rice − 2" from the 3rd equation, and then add 2 to both sides, so 2 × rice = 12. 1 packet of rice = 6 units.) *How can we figure out how much the rice weighs?* (Substitute the mass of the rice in the 3rd equation, so flour = 4 units.)

3 Together, read the question on the blackline master and then ask the students to complete the sheet.

[Equivalence and Equations]

Making Connections

Name _____

1. A farmer wants to pump water from a tank to a house. The total length of pipe he needs is 24 m. There are 3 different lengths of pipe.

Here are 3 ways to make 24 m without cutting the pipes.

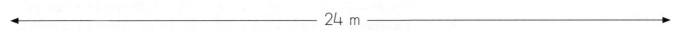

a. Write an equation for each connection above. Then calculate the pipe lengths.

A _____ = 24   Pipe **A** = _____ m

B _____ = 24   Pipe **B** = _____ m

C _____ = 24   Pipe **C** = _____ m

b. Write how you figured it out. _____
_____
_____

2. The distance from a house to another tank is 32 m. Write 6 different ways to connect the tank to the house without cutting any of the pipes above. Try writing each way as an equation.

_____   _____

_____   _____

_____   _____

Mystery Masses

Identifying the relationships between unknowns

AIM

Students will use visual clues to determine relationships between collections of objects. They will then use these relationships to balance equations.

MATERIALS

- 1 copy of the blackline master (opposite) for each student

REFLECTION

Refer to Questions 2 and 3 on the blackline master and discuss the different methods the students used to figure out the unknown. List the methods on the board. For each question, ask, *Is there another way to figure out the unknown?*

1 Refer to Question 1 on the blackline master. Ask volunteers to read the clues. Discuss the clues and then ask, *How can we balance the sharpener and the pencil?* (Add another sharpener to the right side.) Encourage the students to share and explain their answers.

2 Together, work through one solution. Look at the 1st clue and write **2 × eraser = 2 × sharpener** on the board. Say, *So, 1 eraser balances 1 sharpener.* Look at the 2nd clue and write **3 × eraser = 1 pencil + 1 sharpener** on the board. Say, *So, 3 sharpeners balance 1 pencil plus 1 sharpener.* Then, *2 sharpeners balance 1 pencil.*

3 Have the students complete the blackline master. Ask volunteers to share and explain their solutions.

[Equivalence and Equations]

Mystery Masses

Name _____

For each of these, draw an object in the bag to make the picture balance. Use the clues to help you.

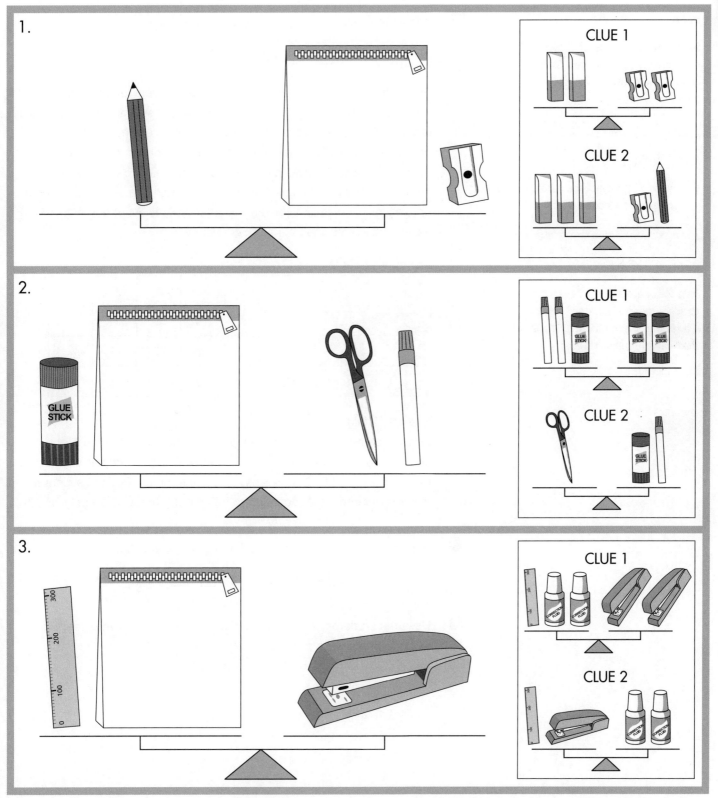

Boxed In

Solving for unknowns in equations that involve addition, subtraction, and multiplication

AIM

Students will replace unknowns with numbers in systems of equations and identify the relationships between the unknowns.

MATERIALS

- 1 copy of the blackline master (opposite) for each student

REFLECTION

Ask, *What are some different ways to solve these equations? Which way helps you the most? Why is that way the most helpful?*

1 Write the following pair of equations on the board.

$$\square + \bigcirc = 45 \qquad \square - \bigcirc = 11$$

Ask, *What do you know about the missing numbers in these equations?* Encourage responses such as, "They are both less than 45; one number is odd and the other is even; they are more than ten apart so they cannot both have the same digit in the tens place; or one number is probably in the 20s." Invite individuals to suggest strategies to find the unknown numbers. For example, guess and check; construct a table; use algebra; or use logical reasoning like the responses above. Methods that involve using a table and algebra are shown below.

Example 1 (a table)

$\square$	32	31	30	29	28	
$\bigcirc$	13	14	15	16	17	
$\square + \bigcirc = 45$	45	45	45	45	45	
$\square - \bigcirc = 11$	19	17	15	13	11	

Example 2 (algebra)

Add the same amounts from the 2nd equation to each side of the 1st equation to get $\square + \bigcirc + \square - \bigcirc = 45 + 11$. Then add and subtract to make $\square + \square = 56$. That means $\square$ is half of 56 or 28. Then substitute this answer into either equation to figure out that the value of $\bigcirc$ is 17.

2 Repeat the discussion for the following pair of equations.

$$\square + \bigcirc = 36 \qquad 2 \times \square - \bigcirc = 24$$

3 Have the students complete the blackline master. Ask volunteers to share their answers.

[Equivalence and Equations]

Boxed In

Name _____

A store sells different-shaped boxes. Write prices in the shapes to make the equations true.

1.

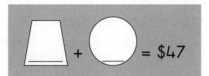 

Write or draw how you figured out the cost of these boxes.

_____
_____
_____
_____

2.

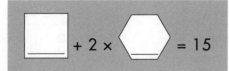 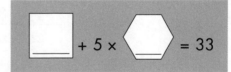

Write or draw how you figured out the cost of these boxes.

_____
_____
_____
_____

Just Juice

Using repeating patterns to introduce ratios

AIM

Students will represent ratios with 2 components as repeating patterns and tables of values. They will also compare parts of ratios to identify patterns.

MATERIALS

- 1 copy of the blackline master (opposite) for each student

REFLECTION

Ask, *If we have 100 repeats, how many pears do we use?* (100) *How many apples?* (300) *Does the juice from the 100th glass taste the same or different to the juice from the 1st glass? Why?* (The same because in both glasses there is 3 times as much apple juice as pear juice.)

1 Discuss fruit juices the students like, and whether they make or buy their juice. Say, *Imagine we make one glass of apple and pear juice. We will use 1 pear and 3 apples. If we make 2 glasses of the same juice, how many pears and how many apples will we need?* (2 pears and 6 apples.) Draw the ingredients, as shown below, on the board.

Ask, *What type of pattern is this?* (A repeating pattern.)

2 Together, complete the first 3 rows of Question 1 of the blackline master, and then ask the students to complete the question individually. Ask the students to discuss the patterns in the table with a partner. Distinguish between looking for patterns across the table and down the table.

Then ask the students to complete Question 2 individually. At the same time, draw on the board 2 copies of the simplified table from Question 1, as shown below (the black text only at this time).

Pears		Apples
1	for every	3
2	for every	6

×3 (across top and bottom)

Pears		Apples
1	for every	3
2	for every	6

÷3 (across top and bottom)

3 Ask volunteers to share their answers to Question 2, and discuss how the number of apples is 3 times the number of pears in every repeat. Ask, *What is another way of describing this pattern? If we know the number of apples but not the number of pears, what rule can we use to figure out the number of pears?* (The number of pears is one-third of the number of apples.) Write the two rules on your tables, as shown above. Ask the students to complete Question 3.

[Patterns and Functions]

Just Juice

Name _____

1. A juice bar makes a glass of fruit punch by mixing the juice of 1 pear with the juice of 3 apples. Complete the table below.

Number of glasses (repeats)	Number of pears	Number of apples
1		
2		
3		
4		
5		

2. Look at the patterns across the table above.

   a. Compare the number of pears to the number of repeats. Write what you notice.

   _____

   b. Compare the number of apples to the number of repeats. Write what you notice.

   _____

   c. Compare the number of apples to the number of pears. Write what you notice.

   _____

3. Complete the table below to show greater quantities of the same recipe.

Number of glasses (repeats)	Number of pears	Number of apples
		21
	9	
		36
	16	
		60

Algebra For All, Red Level — [Blackline Master | Just Juice]

Juicy Juice

Representing ratios with two components as repeating patterns and tables of values, and generating equivalent ratios

AIM

Students will represent ratios with 2 components as repeating patterns and tables of values, and then generate equivalent ratios for different situations.

MATERIALS

- 1 copy of the blackline master (opposite) for each student

REFLECTION

Have the students make their own juice recipe by mixing 2 kinds of juice. Ask them to write recipes for 1 glass (1 repeat), 10 glasses, and 25 glasses. Students can then ask a partner to check their recipes.

1 Say, *A juice bar makes tropical juice by combining glasses of orange and apple juice. For every 1 orange they use 4 apples. For 2 glasses, how many oranges does the juice bar use?* (2) *How many apples?* (8) *How did you figure it out?* (Multiplied the quantity of each ingredient by 2.)

2 Ask the students to complete Questions 1 and 2 on the blackline master. Ask volunteers to share their answers, and then discuss how the number of apples is always 4 times the number of oranges.

3 Say, *The juice bar tries a new recipe for orange and apple juice. They will use 6 oranges and 18 apples. Will this juice taste the same as the first juice?* (No.) Allow the students time to discuss their ideas with a partner before inviting volunteers to share their answers. Ensure they understand that the across pattern for the new fruit juice is different as the number of apples is only 3 times the number of oranges. Have the students complete Question 3. Ask volunteers to share their answers, and discuss how, when making repeats, the numbers of oranges and apples are always multiplied by the same numbers from the 1st repeat.

4 Write **6 oranges for every 18 apples** on the board. Say, *The new recipe is used to make some glasses of juice. If the juice bar wants to make double that amount of juice, how many oranges will they need?* (12) *How many apples?* (36) *If they want to make a smaller amount of juice, what is the smallest amount they can make so that the taste is the same?* Allow the students time to discuss their answers with a partner before inviting volunteers to share their answers. Ensure they understand that a single glass will use 1 orange for every 3 apples. Use the diagrams shown below to aid discussion.

Oranges		Apples
	for every	
6	for every	18
12	for every	36

×2 (from row 2 to row 3) ×2

Oranges		Apples
1	for every	3
6	for every	18
12	for every	36

÷6 (from row 2 to row 1) ÷6

[Patterns and Functions]

Juicy Juice

Name _____

1. A juice bar makes a glass of tropical juice by mixing the juice of 1 orange with the juice of 4 apples. Complete the table below.

Number of glasses (repeats)	Number of oranges	Number of apples
1		
2		
3		
4		
5		

2. Look at the patterns across the table. Write a rule you can use to figure out the number of apples when you know the number of oranges.

   _____

3. Look at the patterns down the table. Compare the 1st and 2nd repeats.

   a. Write how you can figure out the number of oranges in the 2nd repeat.

   _____

   b. Write how you can figure out the number of apples in the 2nd repeat.

   _____

4. Look at the patterns down the table. Compare the 1st and 3rd repeats.

   a. Write how you can figure out the number of oranges in the 3rd repeat.

   _____

   b. Write how you can figure out the number of apples in the 3rd repeat.

   _____

Number Rectangles

Investigating growing patterns of squares and oblong shapes

AIM

Students will develop rules for growing patterns involving rectangular shapes

MATERIALS

- 1 copy of the blackline master (opposite) for each student

REFLECTION

Reinforce the idea that the numbers in both sequences (squares and oblongs) can be determined by a rule based on the position of each rectangle in the sequence.

1 Complete the blackline master for this activity either before or after the lesson.

2 Draw the sequence of oblong shapes and the table shown below on the board. At this stage, only include the numbers in the first row. (*Note:* An oblong is a rectangle that is not a square.)

Picture number	1	2	3	4	5
Number of dots	2	6	12	20	30

Say, *Look at the oblongs. How do the oblongs change? What parts increase?* (Each oblong has one extra row and one additional dot in each row.) *What should we write in the first three empty spaces in the table? What will the next oblong in the pattern look like? What should we write in the table?* Continue for the 5th oblong in the pattern. During the discussion, highlight that the picture number is the same as the number of rows, and that the number in each row is one more than the picture number. Encourage the students to use this observation to predict the next 2 or 3 entries in the table. The rule to determine the number of dots in any oblong in this sequence is "Picture number × (picture number +1)".

3 If they have not already done so, allow time for the students to complete the blackline master. Ask volunteers to share their answers to each question.

Number Rectangles   Name _____

1. a. Extend this growing pattern. Write the number of squares under each picture.

   Picture 1   Picture 2   Picture 3   Picture 4   Picture 5

   ____   ____   ____   ____   ____

   b. Write what you think Picture 15 will look like. _____
   _____

   c. What is the position of Picture 5 in this growing pattern? _____

   d. Write a rule for figuring out the square number in any position. _____
   _____

2. a. Extend this growing pattern to show how square numbers can be created using odd numbers.

   Picture 1   Picture 2   Picture 3   Picture 4   Picture 5

   1   1 + 3   1 + 3 + 5   _____   _____

   (1 × 1)   (2 × 2)   (3 × 3)   (_____)   (_____)

   b. What is the quick way to calculate the sum of 1 + 3 + 5 + 7 + 9 + 11?
   _____

   c. What is the sum of the first

      • 12 odd numbers? _____   • 20 odd numbers? _____

Algebra For All, Red Level   [Blackline Master | Number Rectangles]   27

Frieze Frame

Creating rules that relate objects in patterns

AIM

Students will represent patterns as tables of values and write the rule for each pattern.

MATERIALS

- Pattern blocks (optional)
- 1 copy of the blackline master (opposite) for each student

TEACHING NOTE

In North America, the term "trapezoid" is used to describe a quadrilateral that has only two parallel sides. In Australia, Europe, and elsewhere, the same shape is called a "trapezium".

REFLECTION

Refer to the blackline master and lead a discussion by asking questions such as, *How is the number of triangles (rhombuses) related to the number of hexagons? What rules can we write?*

1 Draw the pattern shown below on the board (or create the pattern using pattern blocks).

Say, *Imagine we are tiling a bathroom. This is our pattern. We always start and end with a triangle.* Draw the table shown below on the board.

Picture number	1	2	3	4	5
Number of trapezoids	1	2	3	4	5
Number of triangles					

Ask, *If we have 1 trapezoid, how many triangles do we have?* (2) *If we have 2 trapezoids, how many triangles do we have?* (4) Complete the table and ask, *If we have 10 (20, 100) trapezoids, how many triangles do we have?* (20, 40, 200) *What rule can we write to determine the number of triangles for any number of trapezoids?* Write **Number of triangles = 2 × Number of trapezoids** on the board.

2 Have the students complete the blackline master. Ask volunteers to share their answers. Refer to Question 1 and ask, *How did you figure out the rule? If we multiply the number of hexagons by 2, what must we do to this number to give us the number of triangles?* Discuss ways of writing the rules including, "Number of triangles = 2 × number of hexagons − 2" and "Number of hexagons = (number of triangles + 2) ÷ 2".

[Patterns and Functions]

Frieze Frame

Name _____

1. A bathroom mirror is framed with a pattern of hexagonal and triangular tiles.

   a. Extend this pattern. Finish each picture with a hexagon.

Picture 1	Picture 2	Picture 3	Picture 4	Picture 5

   b. Complete this table to match the pattern.

Picture number	1	2	3	4	5		
Number of hexagons							
Number of triangles							

   c. If the frieze has 12 hexagons, how many triangles does it have? _____

   d. If the frieze has 46 triangles, how many hexagons does it have? _____

   e. Write a rule for figuring out the number of triangles when the number of hexagons is known.
   _____

2. a. Look at the first 3 pictures in this pattern.

Picture 1	Picture 2	Picture 3
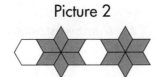		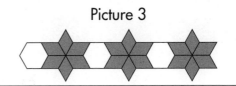

   b. Complete this table to match the pattern.

Picture number	1	2	3	4	5		
Number of hexagons							
Number of rhombuses							

   c. Write a rule for figuring out the number of rhombuses when the number of hexagons is known.
   _____

Bulk Buys

Using multiplication and addition to create rules that are represented using equations or arrow diagrams

AIM

Students will create functions that involve two operations.

MATERIALS

- 1 copy of the blackline master (opposite) for each student

REFLECTION

Refer to the blackline master and discuss the order used for the operations. Ensure the students know what would happen if the order was reversed.

1 Say, *The entry fee to a book fair is $5. At this book fair, the price of each book is $6. What is the total cost of attending this book fair and buying 8 books? What operations should we use? What steps should we follow to figure out the total cost?* During the discussion, draw two function machines on the board, as shown below, to reinforce the order of the rules used to determine the total cost. Encourage the students to explain what would happen if the order of the rules on the machines was reversed.

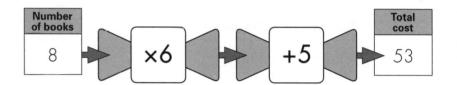

Discuss how the number of books is the IN number and the total cost is the OUT number. Ask, *If we buy 10 books (6 books, 20 books), how much will we spend?* Complete a table to show the total cost for each. Ask, *How can we show this using arrows? What rule can we write to figure out the total cost for any number of books?* Students might use an arrow diagram or equation, as shown below.

$$\text{Number of books} \xrightarrow{\times 6} \xrightarrow{+5} \text{Total cost}$$

or

6 × Number of books + 5 = Total cost

2 Have the students work mentally to complete the blackline master. Call on volunteers to share their answers. Ask, *When is it cheaper to order from the 1st store?* (When ordering more than 3 comics.)

[Patterns and Functions]

Bulk Buys

Name _____

1. a. An online bookstore charges $7 for each of its comics and a flat rate of $1.50 for shipping. Complete this table.

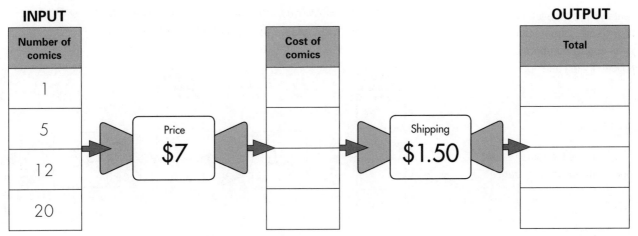

INPUT Number of comics	Cost of comics	OUTPUT Total
1		
5		
12		
20		

b. Use arrows to write a rule for calculating the output number.

_____

c. Write the matching equation. _____

2. a. A different store charges $6 for each of its comics and a flat rate of $4.50 for shipping. Complete this table.

Number of comics	Cost of comics	Total
1		
5		
12		
20		

b. Use arrows to write a rule for calculating the total.

_____

c. Write the matching equation. _____

How Much? How Many?

Backtracking rules that involve two operations

AIM

Students will use a backtrack strategy to calculate input numbers for given output numbers involving multiplication and addition.

MATERIALS

- 1 copy of the blackline master (opposite) for each student

REFLECTION

Encourage the students to explain how they can draw arrows in different directions to show the backtracking. Reinforce the idea that backtracking a rule with two operations means that the inverse operation is used for each step and the steps are followed in reverse order.

1 Draw a diagram with tables and function machines on the board, as shown below. Say, *For each online purchase, Best Buy Books charges $4 per book, plus a flat rate of $5 for shipping.*

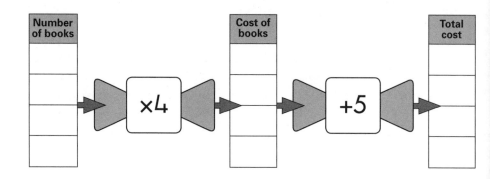

Ask, *How can we show this rule using arrows?* Invite a volunteer to draw the arrows and write the rule on the board, as shown below.

Number of books $\xrightarrow{\times 4}$ $\xrightarrow{+5}$ **Total cost**

2 Ask, *If the total cost is $65, how many books were purchased? How did you figure it out?* (Subtract 5 and then divide by 4.) Introduce the term "backtracking" and say, *When we start with the output and reverse the steps in a rule to find the input, we are backtracking.* Draw the arrow diagram shown below on the board.

Number of books $\xleftarrow{}$ $\xleftarrow{}$ **Total cost**

Ask, *What is the inverse of +5? What is the inverse of ×4? What inverse operation should we do first? How do you know?* Encourage students to explain their thinking and write the inverse operations, **−5** and **÷4**, above the correct arrows.

3 Have the students complete the blackline master. Ask volunteers to share their answers.

[Patterns and Functions]

How Much? How Many?   Name _____

1. a. An online bookstore charges $8 for each of its magazines. It also charges a flat rate of $4.50 for shipping. Complete this table.

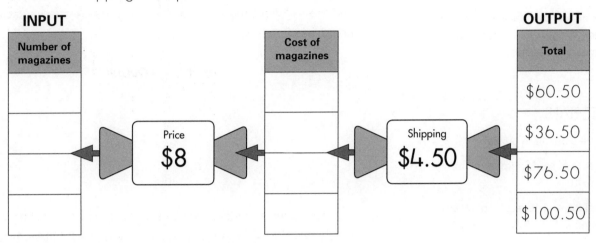

INPUT Number of magazines	Cost of magazines	OUTPUT Total
		$60.50
		$36.50
		$76.50
		$100.50

b. Use arrows to write a rule for calculating the input number.

_____

c. Write the matching equation. _____

2. a. The same bookstore offers a $2.50 discount off the total cost to customers who come in and buy magazines direct from the store. Complete this table.

Number of magazines	Cost of magazines	Total
	$48	$45.50
11		
		$37.50

b. Use arrows to write a rule for calculating the total. Then write the matching equation.

_____

c. Use arrows to write a rule for calculating the number of magazines.
    Then write a matching equation.

_____

What's the Rule?

Determining rules for function machines

AIM

Students will use multiplication and addition to figure out missing rules for function machines.

MATERIALS

- 2 medium-sized boxes
- 2 sets of cards numbered from 0 to 10
- 4 signs — 1 for each operation symbol: "+", "−", "×", and "÷"
- 1 copy of the blackline master (opposite) for each student

REFLECTION

Discuss the importance of the order of the rules. For example, write **× 2** and **+ 4** on the board and complete some example IN numbers and OUT numbers. Invite volunteers to identify the OUT numbers for when the rules are reversed; that is, + 4 first, then × 2. Compare the result with what happens if there are 2 addition rules or 2 multiplication rules.

1 Without being observed, place the "×" sign and a "2" card in the 1st box, and the "+" sign and a "4" card in the 2nd box. Draw a table on the board, as shown below.

Input	Output

Invite 2 students to stand side by side out the front to act as the function machines. Show the boxes and say, *These are 2 function machines. We will figure out the operation and a number for each.*

2 Ask volunteers to write IN numbers in the table on the board. For each suggestion, instruct the 1st student/function machine to mentally multiply the number by 2 and then whisper the result to the 2nd student/function machine. This student then mentally adds 4 and writes the OUT number in the table on the board. After each OUT number is written, ask the students to figure out the rule in each box. As they share their ideas, invite another student to write them on the board beside the table. Repeat until the class successfully figures out the 2 rules.

3 Repeat the activity using different numbers and operations in the function boxes, and different students in each role.

4 Have the students complete the blackline master. Ask volunteers to share their answers.

34 [Patterns and Functions]

What's the Rule?   Name _____

1. a. Write the function on each machine.

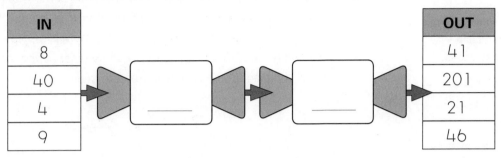

IN		OUT
8		41
40		201
4		21
9		46

   b. Write the complete rule using arrows. _____

   c. Write the matching equation. _____

2. a. Write the function on each machine.

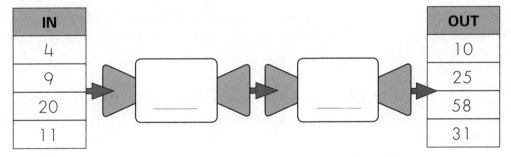

IN		OUT
4		10
9		25
20		58
11		31

   b. Write the complete rule using arrows. _____

   c. Write the matching equation. _____

3. a. Write the function on each machine.

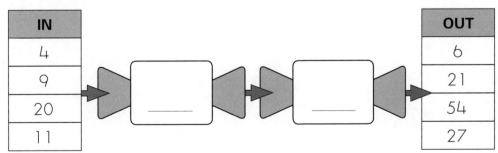

IN		OUT
4		6
9		21
20		54
11		27

   b. Write the complete rule using arrows. _____

   c. Write the matching equation. _____

Forward, Backward

Using arrow diagrams to work backward to solve real-world problems

AIM

Students will represent problems with two operations to use backtracking to find solutions.

MATERIALS

- 1 copy of the blackline master (opposite) for each student

REFLECTION

Ask, *If we know the IN number, and the rule is "multiply by 10, subtract 5 and divide by 7", how can we show this using an arrow diagram? If we know the OUT number, how can we figure out the IN number?* Together, create another money problem using all 4 operations. Write some OUT numbers on the board and encourage the students to backtrack to find the IN numbers.

1 Show a wallet to the students and say, *I had some money in my wallet. I went to the store and spent half my money on books. Then my sister gave me another $4. I now have $25 in my wallet. How much money did I start with?* ($42) *How can we show what happened using arrows?* Draw the diagram shown below on the board.

$$\boxed{?} \xrightarrow{\div 2} \xrightarrow{+4} 25$$

2 Ask, *How can we figure it out?* (Reverse the process or backtrack.) Draw the diagram shown below on the board.

$$\boxed{?} \xleftarrow{\times 2} \xleftarrow{-4} 25$$

Discuss how subtraction is the inverse of addition, and multiplication is the inverse of division. Check that the answer is correct by dividing 42 by 2 and then adding 4.

3 Have the students complete the blackline master. Ask volunteers to share their answers.

[Patterns and Functions]

Forward, Backward

Name _____

1. Three classes each raised the same amount of money for charity. The school donated an extra $300 to increase the total to $750.

   a. Write the missing numbers to complete the rule and figure out how much each class raised.

   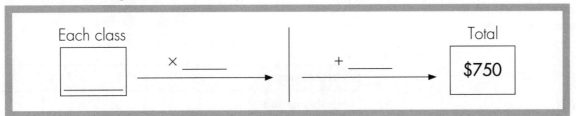

   b. Reverse the arrows and operations to show how the amount raised by each class was calculated.

   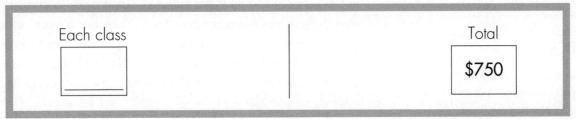

2. For each of these stories, choose a method from Question 1 to help you calculate the missing amount.

   a. Jacob bought airline tickets for his family. Tax and charges on each ticket was $50. He bought 3 tickets. What was the cost of each ticket?

   b. The owners of a new restaurant are planning to have 4 chairs at each table. They also want 5 spare chairs. If they buy 57 chairs, how many tables will they need?

Odds or Evens

Using rules to find or describe odd and even numbers

AIM

Students will find a particular odd or even number and backtrack to identify the position of odd or even numbers.

MATERIALS

- 1 copy of the blackline master (opposite) for each student

REFLECTION

Refer to the blackline master and ask, *If we find the odd number by multiplying its position by 2 and adding 1, how do we find the position if we know the odd number? What is the inverse of multiplying by 2 and adding 1?* (Subtracting 1 and dividing by 2.)

1 Draw the table and sequence of pictures shown below on the board.

Picture number	1	2	3		
Number of dots	2	4	6		

Ask, *What will the next picture look like? What numbers should we write in the table?* (4 and 8.) *What numbers come next?* (5 and 10.) Discuss how these are even numbers and each "number of dots" is double the "picture number". Invite students to suggest a rule that can be written as an equation or shown with an arrow diagram to figure out the 100th (250th) even number. (The 100th even number = 100 × 2; the 250th even number = 250 × 2.)

2 Say, *We know 360 is an even number. What is the position of 360 in the sequence of even numbers?* (180th even number.) *How do you know?* Invite individuals to explain how they backtracked to determine the position.

3 Ask the students to complete the blackline master. Encourage students to describe how they figured out the rule and backtracked the operations.

[Patterns and Functions]

Odds or Evens

Name _____

1. a. Draw more dots to extend this growing pattern.

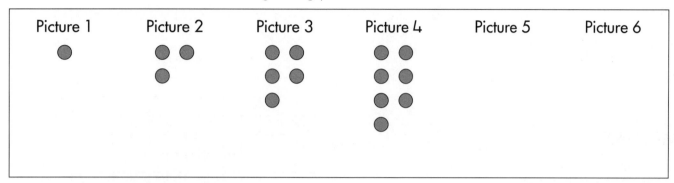

b. Complete this table.

Picture number	1	2	3	4	5	6		
Number of dots								

c. How many dots will be in Picture 15? _____

Write how you know. _____

_____

d. Write the missing numbers to complete a rule that can be used to figure out the number of dots in any picture.

e. Backtrack the rule above to figure out the position of each of these odd numbers in the sequence.

| 49 is the _____th odd number. | 99 is the _____th odd number. |
| 199 is the _____th odd number. | 279 is the _____th odd number. |

Partial Products

Investigating the distributive property for multiplication over addition

AIM

Students will use the distributive property to simplify situations involving multiplication.

MATERIALS

- 1 copy of the blackline master (opposite) for each student

REFLECTION

Write **12 × 9 = __** on the board. Ask, *How else can we write this equation to make it easier to figure out?* Invite individuals to show their thinking using equations such as 12 × 9 = 12 × (5 + 4) = 12 × 5 + 12 × 4 = 108; 12 × 9 = (10 + 2) × 9 = 10 × 9 + 2 × 9 = 108, or 12 × 9 = 12 × (10 − 1) = 12 × 10 − 12 × 1 = 108. Discuss the idea that if you multiply the sum of 2 numbers by another number, you can also multiply each added number individually and then add the products.

1 Draw a 7-by-12 array on the board and label it as shown below.

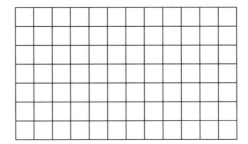

Say, *Imagine these are the seats for the class concert. If all the seats are filled, how many people are attending the concert? How can we figure out the answer (product)?* Invite volunteers to suggest how they can split one factor into two parts to simplify the process. Encourage them to shade the array to show their thinking. Together, write equations on the board to show their thinking. For example:

Total number of people = 7 × 12 = 7 × (10 + 2) = 7 × 10 + 7 × 2 = 84

or

7 × 12 = (5 + 2) × 12 = 5 × 12 + 2 × 12 = 84

2 Repeat the discussion for a 6-by-15 array.

3 Have the students complete the blackline master. Ask volunteers to share their answers.

Partial Products

Name _____

1. Follow these steps:
   - Shade green the region that shows the tens. Then write the product.
   - Shade red the region that shows the ones. Then write the product.
   - Write the total of the 2 partial products.

a. 6 × 10 = _____ 6 × 2 = _____	b. 4 × 20 = _____ 4 × 3 = _____

6 × 12 = _____	4 × 23 = _____
c. 5 × 20 = _____ 5 × 6 = _____	d. 8 × 10 = _____ 8 × 5 = _____
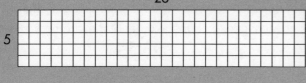	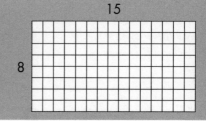
5 × 26 = _____	8 × 15 = _____

2. Split one factor to show 2 parts. Shade the 2 parts and write the product of each. Then write the total of the 2 partial products.

a.	b.	c.
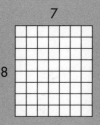	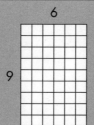	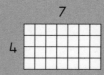
___ × ___ = ___	___ × ___ = ___	___ × ___ = ___
___ × ___ = ___	___ × ___ = ___	___ × ___ = ___
8 × 7 = _____	9 × 6 = _____	4 × 7 = _____

Algebra For All, Red Level

Look Both Ways

Investigating the commutative property that involves multiplying decimal fractions

AIM

Students will discover that changing the order when multiplying decimal fractions does not change the answer.

MATERIALS

- 1 copy of the blackline master (opposite) for each student

REFLECTION

Discuss the strategies the students used to figure out the answers on the blackline master. Ask, *Does it matter what order we multiply 2 numbers in? Will the answers be the same or different?*

1 Sketch and label the garden plot shown below left on a large sheet of paper.

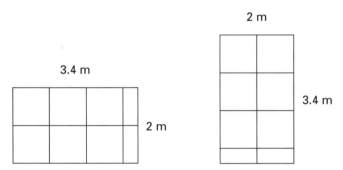

Show the sheet of paper to the students, orientating the garden plot as shown above left. Ask, *What do you know about the area of this garden?* Encourage the students to describe the area as accurately as possible using strategies of their choice. Turn the sheet of paper as shown in the second picture and ask, *What is the same (different) about this picture? What equation can we write for each picture?* Highlight that the area has not changed, though the order of the numbers in each equation is different.

2 Have the students work mentally to complete the blackline master. Ask volunteers to share their answers.

[Properties]

Look Both Ways

Name _____

1. Write 2 turnaround multiplication facts that can be used to figure out the area of each grid.

a.
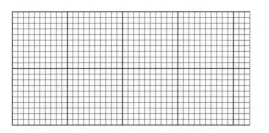

2 × _____ = _____   _____ × 2 = _____

b.
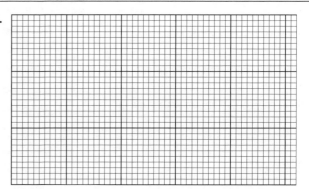

3 × _____ = _____   _____ × 3 = _____

c.
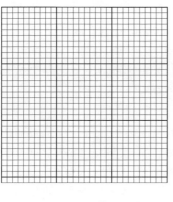

_____ × 3 = _____   3 × _____ = _____

d.
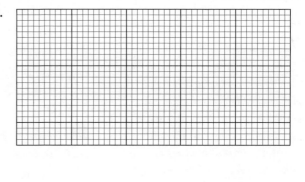

_____ × 5 = _____   5 × _____ = _____

2. Shade the grid to show the expression. Then write 2 turnaround multiplication facts to match.

a. 2 × 3.6

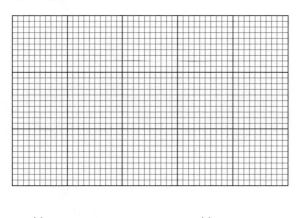

_____ × _____ = _____   _____ × _____ = _____

b. 2.5 × 4

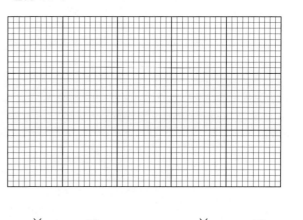

_____ × _____ = _____   _____ × _____ = _____

Algebra For All, Red Level   [Blackline Master | Look Both Ways]

Rearranging Rectangles

Adjusting the factors in a multiplication number sentence to help figure out the answer

AIM

Students will discover that multiplying one factor by a number in a multiplication number sentence, and dividing the other factor by the same number, does not change the answer.

MATERIALS

- 1 copy of the blackline master (opposite) for each pair of students
- Scissors and tape

REFLECTION

Discuss how these methods help simplify multiplication expressions to make them mentally manageable. Ask, *If we divided one factor by 3 to simplify it, what should we do to the other factor so the answer stays the same?*

1 Arrange the students into pairs. Provide each pair with a copy of the blackline master. Read the instructions and ask each pair of students to work together to cut out the first array. Then ask, *How can you cut this rectangle so that when you rearrange the pieces, you make a new array?* After the students form a new array, discuss the new array and write the dimensions of both rectangles on the board as an equation ($3 \times 27 = 9 \times 9$). Discuss questions such as, *How do we know that the two equations are equal? How did the array change?* (One dimension or factor was tripled while the other factor is one-third as great.) *Which side of the equation is easier to figure out? Why?*

2 Repeat the steps for the second array. Encourage the students to find all the ways this rectangle can be rearranged as an array and write each expression. Use the steps above to discuss the answers and write all the possible equations. For each equation, reinforce the fact that one factor has been divided by a number while the other factor has been multiplied by the same number.

3 Ask students to work in groups to write every possible equation for the other arrays on the blackline master. Have them try to write the equations without cutting and rearranging the arrays. Ask volunteers to share their answers.

[Properties]

Rearranging Rectangles

Name _____

Follow these steps:

1. Cut out Rectangle **A**.
2. Cut the rectangle into a number of equal parts.
3. Rearrange the pieces to make a new rectangle.
4. Record the dimensions.
5. Repeat Steps 2 to 4 for Rectangles **B**, **C**, and **D**.

A 3 × 27

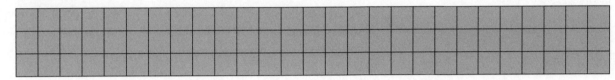

B 3 × 24

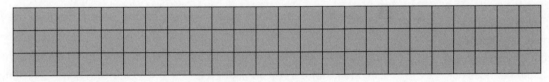

C 6 × 15

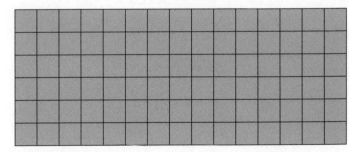

D 4 × 18

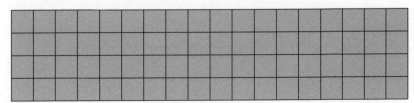

No Change

Investigating the effect of multiplying and dividing by the same number

AIM

Students will learn that multiplying and dividing by the same number is like multiplying by 1.

MATERIALS

- 1 copy of the blackline master (opposite) for each student

REFLECTION

Explain that multiplying and dividing by the same number leaves the starting number unchanged, and is the same as multiplying (or dividing) the number by 1. *Note:* Because multiplying a number by 1 does not change the number (in the same way that adding 0 to a number does not change a number), 1 is called the "identity element" for multiplication.

1 Draw the arrow diagram shown below on the board.

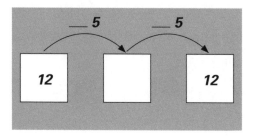

Ask, *What operations can we use with the 5 on each arrow so we start and end with 12?* Invite individuals to suggest the combinations of addition/subtraction and multiplication/division in different orders.

2 Have the students complete the blackline master.

3 Write the equations shown below on the board. Point to the first equation and ask, *What is the missing number? How do you know?* Encourage the students to explain their thinking. Then reinforce the idea that expressions such as "× 2 ÷ 2" are the same as multiplying by 1 because they leave the number unchanged.

$$23 \times 2 \div 2 = \square$$

$$\square \times 3 \div 3 = 17$$

$$19 \times \square \div 5 = 19$$

$$31 \times 4 \div \square = 31$$

4 Repeat the discussion for the other number sentences.

46 [Properties]

No Change

Name _____

1. For each of these, write **+**, **−**, **×**, and **÷** to make 4 different arrow diagrams that are true. Write the missing numbers.

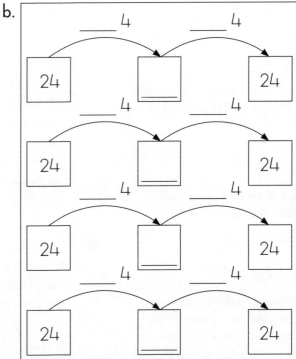

2. For each of these, use **×** and **÷** to complete 2 different arrow diagrams that are true. Write all the missing numbers. There is more than one answer.

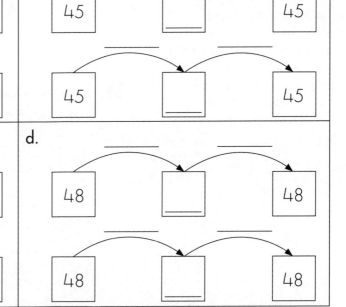

Algebra For All, Red Level — [Blackline Master | No Change]

Following Orders

Using brackets or parentheses ( ) to indicate the order of operations

AIM

Students will use a real-world situation to investigate why the order in which they apply operations is important. They will then use brackets or parentheses ( ) to show the order that should be followed.

MATERIALS

- 1 copy of the blackline master (opposite) for each student

TEACHING NOTE

The examples in this lesson were chosen because many international studies show that students tend to figure out 28 − 5 + 3 by adding the 5 and 3 and then subtracting 8 from 28. They also add before multiplying in the example 9 + 5 × 12.

REFLECTION

Discuss the order in which to perform operations. Reinforce that the operation in brackets ( ) is done first. Multiplication and division are completed before addition and subtraction. Operations at the same level are completed working left to right.

1 Write the equation and prices shown below on the board.

28 − 8 + 3 = ___   **$28**   **$8**   **$3**

Ask, *What is the answer? How do you know?* Invite students to create stories for the equation using the 3 amounts of money, for example, "Jo had $28. She spent $8 and was then given $3." Highlight the fact that subtracting and then adding gives a different answer to adding and then subtracting. Explain that brackets ( ) help indicate the order to follow. Further explain that if there are no brackets, and the operations are "at the same level" (for example, addition and subtraction or multiplication and division), then the answer is calculated by working left to right.

2 Read Question 1 on the blackline master with the class. Say, *Imagine that you buy 1 CD and 5 DVDs. How much will you pay? What equation matches the thinking you used?* Invite individuals to describe their thinking and write equations. Discuss equation examples, such as 9 + 5 × 12 = ___ or 5 × 12 + 9 = ___, to show that working left to right will give different answers and that the first answer does not make sense. Reinforce that multiplication is performed before addition (or subtraction).

3 Repeat the discussion, saying, *Mary buys 9 DVDs and Meg buys 5 DVDs. What will they pay in total?* For this situation, discuss how the equations below can be used to show the thinking.

(9 + 5) × 12 = ___

or

9 × 12 + 5 × 12 = ___

4 Have the students complete the blackline master. Ask volunteers to share their answers.

[Properties]

Following Orders

Name _____

Digital Sale

Music CDs  $9 each

DVDs  $12 each

eBooks $7 each

Games  $15 each

1. Write an equation to show how you would figure out the total cost of each purchase. Then write the answer.

a.	1 DVD and 7 eBooks _____	b.	5 eBooks and 1 CD _____
c.	4 Games and 4 DVDs _____	d.	6 CDs and 1 DVD _____
e.	8 eBooks and 6 Games _____	f.	1 eBook and 5 CDs _____
g.	7 CDs and 1 Game _____	h.	3 CDs and 3 eBooks _____

2. Write the answers to these equations.

   a. $16 + 8 - 4 =$ _____

   b. $29 - 6 + 11 =$ _____

   c. $8 + 36 \div 4 =$ _____

   d. $3 \times 9 - 4 \times 2 =$ _____

   e. $2 \times 32 \div 8 =$ _____

   f. $48 \div 6 \times 2 =$ _____

Building Blocks

Investigating the associative property for multiplication

AIM

Students will use brackets or parentheses ( ) to describe the associative property and specify the order in which to multiply three or more numbers.

MATERIALS

- 1 copy of the blackline master (opposite) for each student

REFLECTION

Encourage students to describe the associative property in their own words, and the fact that brackets or parentheses ( ) are used to indicate which two numbers should be multiplied first.

1 Sketch the 2-by-4-by-6 rectangular prism shown below on the board.

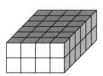

Ask, *How many blocks are in this prism? How do you know? What are some ways we can figure out the total? Does it matter which numbers we multiply first?* Invite the students to suggest the order and use brackets to indicate that the first two numbers are multiplied first.

2 Write ***2 × (4 × 6) = ___*** on the board and ask, *What will you multiply first in this equation? How do you know?* Reinforce the idea that the brackets show what numbers to multiply first. Repeat the discussion for other equations using the numbers 2, 4, and 6.

3 Write ***2 × (4 × 6) = (2 × 4) × 6*** on the board and invite students to explain what is the same and different about each side of the equation. Reinforce the idea that although the numbers are multiplied in different orders, the answers are the same.

4 Have the students complete the blackline master. Ask volunteers to share and explain their answers.

[Properties]

Building Blocks

Name _____

1. These 2 equations describe the dimensions of this prism. For each equation, draw ( ) to show which pair of numbers you would multiply first. Then write the products.

   a. 3 × 5 × 7 = _____

   b. 3 × 7 × 5 = _____

2. Write the missing numbers. Multiply the numbers in the ( ) first.

   a. | (6 × 2) × 4 | 6 × (2 × 4) |
      | = ___ × ___ | = ___ × ___ |
      | = ___ | = ___ |

   b. | (6 × 5) × 3 | 6 × (5 × 3) |
      | = ___ × ___ | = ___ × ___ |
      | = ___ | = ___ |

   c. | (5 × 3) × 8 | 5 × (3 × 8) |
      | = ___ × ___ | = ___ × ___ |
      | = ___ | = ___ |

   d. | (15 × 4) × 5 | 15 × (4 × 5) |
      | = ___ × ___ | = ___ × ___ |
      | = ___ | = ___ |

3. Draw ( ) to show the different pairs you could multiply first. Write the missing numbers.

   a. | 6 × 25 × 4 | 6 × 25 × 4 |
      | = ___ × ___ | = ___ × ___ |
      | = ___ | = ___ |

   b. | 4 × 75 × 2 | 4 × 75 × 2 |
      | = ___ × ___ | = ___ × ___ |
      | = ___ | = ___ |

High in the Sky

Interpreting points on a coordinate graph

AIM

Students will read a table of data, interpret points on a graph, and give reasons for their choices.

MATERIALS

- 1 copy of the blackline master (opposite) for each student

REFLECTION

Refer to the blackline master and ask, *How did you figure out which bird each point represents? Did you focus on a particular point first? Or did you focus on a particular altitude or wingspan? Why did you choose that point? Was it the bird with the longest wingspan or the highest altitude?*

1 Draw a graph with points as shown below on the board.

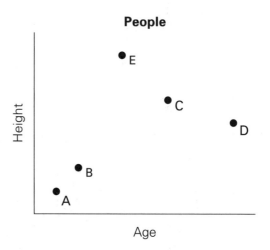

Also draw representations of the following people: a baby, a boy, a grandmother, a father, and a teenage basketballer. These are in order of height, with the baby being the shortest and the basketballer the tallest. Position the drawings randomly, for example: grandmother, boy, teenage basketballer, baby, father.

Ask, *Which point on the graph is the baby? How do you know?* Explain that the further a point is along each axis, the higher its value. Ask, *Which point is furthest up in the height direction? Who do you think this point represents? Which point is furthest along in the age direction? Who do you think this point represents?* On the board, briefly list the students' reasons for naming each point, for example, "Point E is the basketballer. He is the tallest but he is younger than the father and the grandmother".

2 Together, read Question 1 on the blackline master. Ask, *What do we mean by wingspan? Which bird has the longest wingspan? Which bird has the shortest wingspan? Which bird flies at the highest altitude? Which bird flies at the lowest altitude?* Ask the students to complete the blackline master. Call on volunteers to share their answers and explain their selections.

High in the Sky

Name ____________________

Bird	Sparrow	Pigeon	Sandpiper	Goose	Owl	Eagle
Highest altitude (m)	600	4000	3500	1200	1000	1000
Wingspan (cm)	25	46	37	89	112	250

1. Write the name of the bird at each point of the graph.

A ____________________

B ____________________

C ____________________

D ____________________

E ____________________

F ____________________

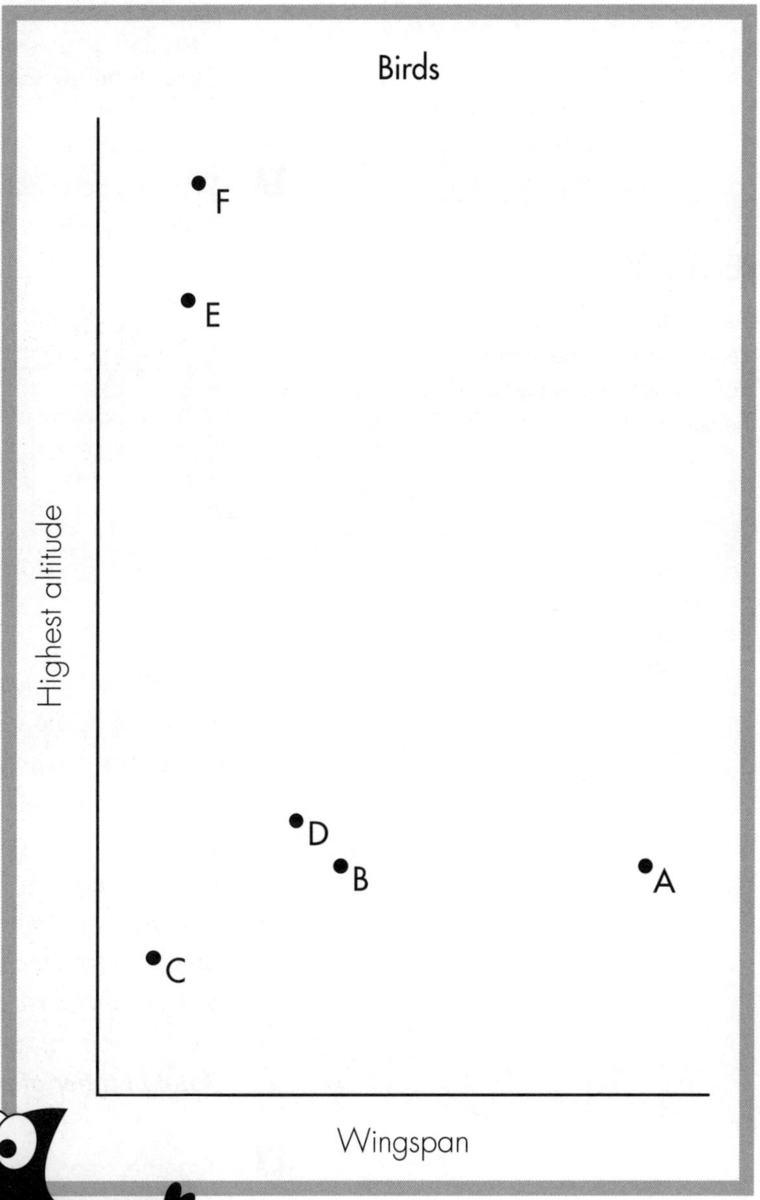

2. What other variable might determine the altitude a bird can reach?

____

____

Card Games

Extrapolating and interpolating in graphs that involve discrete data

AIM

Students will represent simple multiplication and division situations as graphs and use the data to interpolate and extrapolate.

MATERIALS

- 1 copy of the blackline master (opposite) for each student

REFLECTION

Discuss the questions on the blackline master. Reinforce why it does not make sense to join the points.

1 Ask, *If 6 playing cards are shared among 6 people, how many cards will each person receive?* (1) *If there are 18 cards (36 cards), how many will each person receive?* (3, 6) Draw a table on the board, as shown below, and call on volunteers to solve the missing values.

Number of playing cards	6	18	24	48	60
Number of playing cards per person					

2 Draw a graph on the board, as shown below, and ask volunteers to plot the points from the table above.

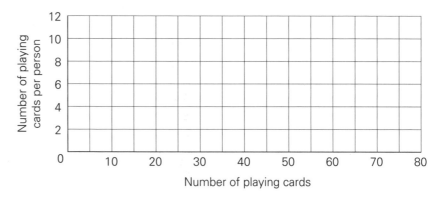

Ask, *Should we join the points? What pattern is in the graph? If there are 42 cards, how many will each person receive? How can we figure this out on the graph?* Using a ruler laid along the existing points, plot all the missing points (12 cards, 30 cards, 36 cards, 42 cards, 54 cards). Ask, *If each person receives 9 cards, what is the total number of cards?* Read the answer from the graph. Ask, *If there are 72 cards in total, how many will each person receive? How can we show this on the graph?* Extend the graph to show 66 cards, 72 cards, and 78 cards. Ask, *What is the rule to figure out the number of cards for each person for any total number of cards?* Write **Number of playing cards per person = Total number of playing cards ÷ 6** on the board.

3 Together, read the blackline master and ask the students to complete the questions. Call on volunteers to share their answers.

[Representations]

Card Games

Name _____

Each player gets 4 cards.

1. Complete the table below to show how many cards are needed.

Number of players	1	2	3	4	5	6	7	8
Number of cards	4							

2. Write how you can figure out the number of cards when you know the number of players.

   _____

3. a. Write the data from the table above as ordered pairs.

   (____,____) (____,____) (____,____)

   (____,____) (____,____) (____,____)

   (____,____) (____,____)

   b. Plot the ordered pairs above on the graph at right.

   c. Write a name for the graph.

   d. Write what you notice about the points on the graph.

   _____

   _____

   _____

   e. Write why it does not make sense to draw a line to connect the points.

   _____

   _____

   _____

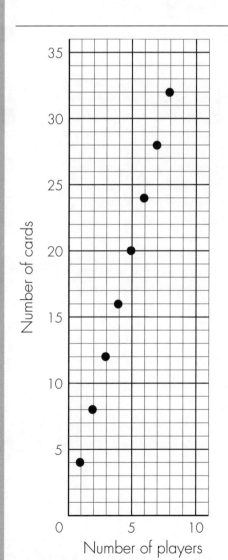

Prize Money

Representing simple division problems as straight-line graphs, and comparing graphs for different rates of change

AIM

Students will represent simple division problems as straight-line graphs. They will also compare graphs for different rates of change.

MATERIALS

- 1 copy of the blackline master (opposite) for each student

REFLECTION

Refer to the blackline master and ask, *Which graph is steeper? Why is it steeper? If we share the money among six people, will the graph be steeper or flatter? Why? If we share the money between 2 people, will the graph be steeper or flatter? Why?* Discuss the idea that the steeper the graph, the fewer people are sharing, and the more money each receives. Ask, *How can you use the graph to figure out the difference between the amount one person receives in each of the different-sized groups?* Discuss how they could figure out the vertical distance between the points on each graph for each amount on the horizontal axis.

1 Together, read Question 1 on the blackline master. Ask, *If three people equally share $1800, how much will each person receive? How did you figure it out?* Ask the students to complete the first row of the table. Ask volunteers to share their answers.

2 Read Question 2 together and say, *We will be drawing three sets of points on this coordinate grid. How can we identify the different sets of points?* (For example, use a different permanent marker for each and write a key at the side of the grid.) Ask the students to plot the points for this set of data. They can then work with a partner to check the positions of the points.

3 Have the students complete the second row of the table and plot the points. They should use a different permanent marker for this set of points.

4 Instruct the students to follow the same steps to plot the third set of points and complete Question 2. Ask volunteers to share their answers.

[Representations]

Prize Money

Name _____

1. A raffle ticket costs $12. There are 6 prize draws. Complete this table to show the amounts of the winnings when shared by 3, 4, or 5 people. Then plot the 3 sets of points on the grid.

Prize money	$300	$600	$900	$1200	$1500	$1800
Shared by 3						
Shared by 4						
Shared by 5						

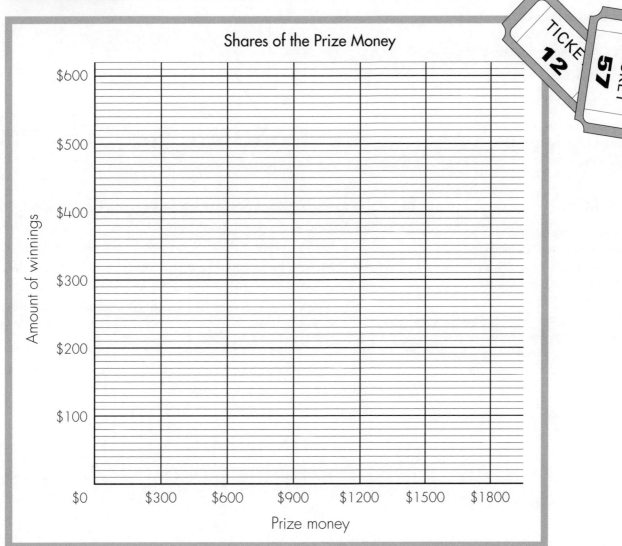

2. a. Which set of data is the steepest?

   _____

   b. Which set of data is the flattest?

   _____

Algebra For All, Red Level — Blackline Master | Prize Money — 57

Fill It Up

Interpreting graphs that show different rates of change

AIM

Students will interpret and construct line graphs related to real-world situations involving time and different rates of change.

MATERIALS

- 2 transparent glass vases: 1 that is a short, wide shape and 1 that is a tall, thin shape
- Access to water
- Funnel and stand (optional)
- 1 copy of the blackline master (opposite) for each student

REFLECTION

Ask volunteers to explain their answers. Discuss how the different slopes of the graph show different rates of change. Say, *Imagine that halfway through filling the vases, the phone rings and you answer it. What will the graphs look like for the time that you are on the phone?* Invite volunteers to sketch their line-graph representations on the board.

1 Show the vases to the students. Ask, *Which is taller?* Slowly fill the tall vase, then the short vase. Pour the water at a constant rate. (*Note:* If the necessary equipment is available, set it up as an experiment with a funnel positioned above the vases.) Ask, *Did the height of the water increase faster in the short vase or the tall vase?* Empty the vases and fill them again, encouraging the students to observe the heights of the water and how fast they change.

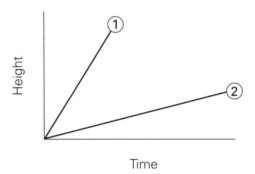

2 Draw the labeled axes of the graph shown above on the board. Draw the 2 lines as shown and label them "1" and "2". Ask, *Which line shows the short vase and which shows the tall vase? How do you know?* Discuss the fact that Line 1 represents a vase that is taller and filling faster because the line is higher in the vertical direction, indicating a taller height, and it is steeper, indicating faster filling.

3 Have the students complete the blackline master. Ask volunteers to share their answers. Ensure that the students relate each component of the graph to the shape of the vase.

[Representations]

Card Games

Name _____

Each player gets 4 cards.

1. Complete the table below to show how many cards are needed.

Number of players	1	2	3	4	5	6	7	8
Number of cards	4							

2. Write how you can figure out the number of cards when you know the number of players.

_____

3. a. Write the data from the table above as ordered pairs.

   (____,____) (____,____) (____,____)

   (____,____) (____,____) (____,____)

   (____,____) (____,____)

   b. Plot the ordered pairs above on the graph at right.

   c. Write a name for the graph.

   d. Write what you notice about the points on the graph.

   _____

   _____

   e. Write why it does not make sense to draw a line to connect the points.

   _____

   _____

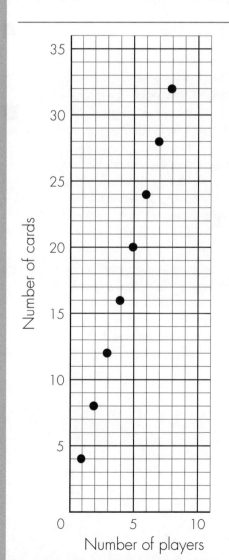

Prize Money

Representing simple division problems as straight-line graphs, and comparing graphs for different rates of change

AIM

Students will represent simple division problems as straight-line graphs. They will also compare graphs for different rates of change.

MATERIALS

- 1 copy of the blackline master (opposite) for each student

REFLECTION

Refer to the blackline master and ask, *Which graph is steeper? Why is it steeper? If we share the money among six people, will the graph be steeper or flatter? Why? If we share the money between 2 people, will the graph be steeper or flatter? Why?* Discuss the idea that the steeper the graph, the fewer people are sharing, and the more money each receives. Ask, *How can you use the graph to figure out the difference between the amount one person receives in each of the different-sized groups?* Discuss how they could figure out the vertical distance between the points on each graph for each amount on the horizontal axis.

1 Together, read Question 1 on the blackline master. Ask, *If three people equally share $1800, how much will each person receive? How did you figure it out?* Ask the students to complete the first row of the table. Ask volunteers to share their answers.

2 Read Question 2 together and say, *We will be drawing three sets of points on this coordinate grid. How can we identify the different sets of points?* (For example, use a different permanent marker for each and write a key at the side of the grid.) Ask the students to plot the points for this set of data. They can then work with a partner to check the positions of the points.

3 Have the students complete the second row of the table and plot the points. They should use a different permanent marker for this set of points.

4 Instruct the students to follow the same steps to plot the third set of points and complete Question 2. Ask volunteers to share their answers.

[Representations]

56

Prize Money

Name _____

1. A raffle ticket costs $12. There are 6 prize draws. Complete this table to show the amounts of the winnings when shared by 3, 4, or 5 people. Then plot the 3 sets of points on the grid.

Prize money	$300	$600	$900	$1200	$1500	$1800
Shared by 3						
Shared by 4						
Shared by 5						

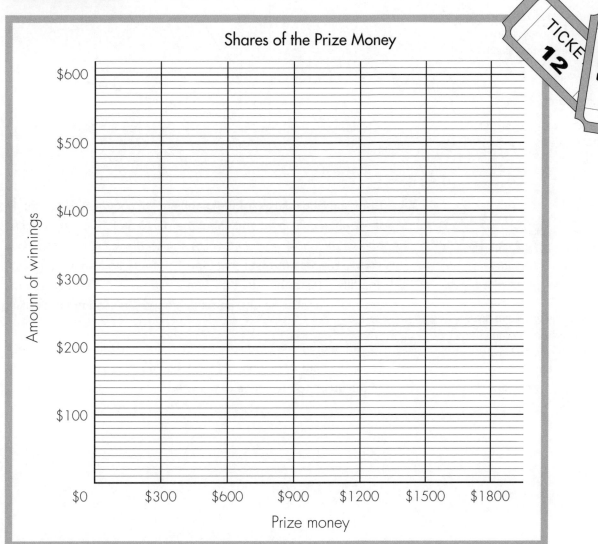

2. a. Which set of data is the steepest?

   _____

   b. Which set of data is the flattest?

   _____

Algebra For All, Red Level   [Blackline Master | Prize Money]   57

Fill It Up

Interpreting graphs that show different rates of change

AIM

Students will interpret and construct line graphs related to real-world situations involving time and different rates of change.

MATERIALS

- 2 transparent glass vases: 1 that is a short, wide shape and 1 that is a tall, thin shape
- Access to water
- Funnel and stand (optional)
- 1 copy of the blackline master (opposite) for each student

REFLECTION

Ask volunteers to explain their answers. Discuss how the different slopes of the graph show different rates of change. Say, *Imagine that halfway through filling the vases, the phone rings and you answer it. What will the graphs look like for the time that you are on the phone?* Invite volunteers to sketch their line-graph representations on the board.

1 Show the vases to the students. Ask, *Which is taller?* Slowly fill the tall vase, then the short vase. Pour the water at a constant rate. (*Note:* If the necessary equipment is available, set it up as an experiment with a funnel positioned above the vases.) Ask, *Did the height of the water increase faster in the short vase or the tall vase?* Empty the vases and fill them again, encouraging the students to observe the heights of the water and how fast they change.

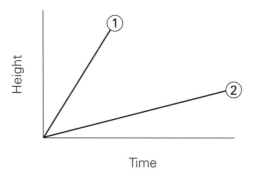

2 Draw the labeled axes of the graph shown above on the board. Draw the 2 lines as shown and label them "1" and "2". Ask, *Which line shows the short vase and which shows the tall vase? How do you know?* Discuss the fact that Line 1 represents a vase that is taller and filling faster because the line is higher in the vertical direction, indicating a taller height, and it is steeper, indicating faster filling.

3 Have the students complete the blackline master. Ask volunteers to share their answers. Ensure that the students relate each component of the graph to the shape of the vase.

[Representations]

Fill It Up

Name _____

1. Each of these glasses were filled at the same rate.

   A  B 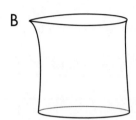 C 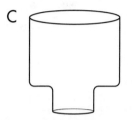 D

   Write **A**, **B**, **C**, and **D** to label each graph with its matching glass.

   a. 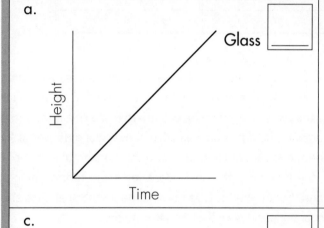 Glass ____

   b. 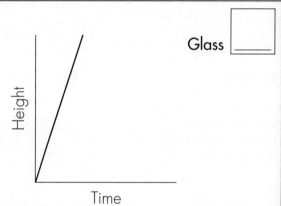 Glass ____

   c. 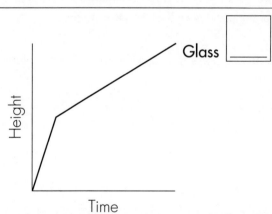 Glass ____

   d. 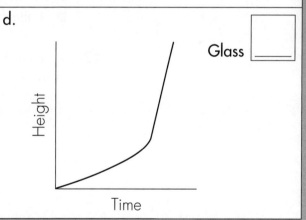 Glass ____

2. Sketch a line on this graph to match the glass below.

   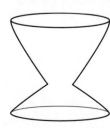

It Depends

Identifying the independent and dependent variable in real-world situations

AIM

Students will label the axes of a graph according to the context of the question by identifying the independent and dependent variable.

MATERIALS

- 1 copy of the blackline master (opposite) for each student

REFLECTION

Ask students to write rules for each of the questions on the blackline master. Invite volunteers to share and explain their rules.

1 Say, *We are buying CDs. They cost $4 each.* Draw a table on the board, as shown below, and complete it with the class.

Number of CDs	Total cost
2	
5	
7	
9	
10	
20	

2 Ask, *What information do we know? What rule can we use to figure out the unknown?* Then discuss how the total cost "depends" on the number of CDs. Draw the axes of a graph on the board and explain that graphs are constructed with the dependent variable shown along the vertical axis and the independent variable shown along the horizontal axis. So, for this situation, the total cost (the dependent variable) is shown along the vertical axis, and the number of CDs (the independent variable) is shown along the horizontal axis. Label the axes accordingly.

3 Have the students complete the blackline master. Ask volunteers to share their answers.

[Representations]

It Depends

Name _____

For each of these, read the story and complete the table. Then label each axis of the graph to match.

1. Jarad priced a pack of 4 CDs at 5 different music stores. What was the cost of each CD in each store?

Total cost	$28	$25	$30	$26	$32
Cost per CD					

2. Jacinta wants to buy a number of CDs. Each CD costs $7.50. How much will she pay?

Number of CDs	1	2	3	4
Total cost				

3. Jack wanted 4 new-release single CDs. He priced them at 3 different stores. What was the total cost in each store?

Cost per CD	$3.50	$4.25	$5.15
Total cost			

4. Jessica made 4 gift-packs of CDs for her friends. Each CD cost $8.00. How many CDs did each friend receive?

Total cost	$32	$24	$48	$56
Number of CDs				

Algebra For All, Red Level

The Great Escape

Determining the appropriate scale for each axis of a graph

AIM

Students will determine and label the dependent and independent variable and determine the appropriate scale for each axis of a coordinate graph.

MATERIALS

- 1 copy of the blackline master (opposite) for each student

REFLECTION

Discuss the information the students wrote along the axes to reinforce that they should show the independent and dependent variables.

1 Tell the students a story about a snail that climbs out of a well. The snail climbs 3 m during the daytime, but slides back 2 m during the night when it is asleep. Ask, *How can we show the height the snail has reached at the end of each daytime period?* Encourage the students to suggest showing this information on a table and a graph. Draw the table shown below on the board and complete it with the class.

End of daytime (days)	1	2	3	4	5	6
Height reached (m)	3	4				

2 Draw the axes of a graph on the board. Ask, *What should we draw and write on a graph to show the same data?* Have the students explain why "End of daytime" is written along the horizontal axis and "Height reached" along the vertical axis of the graph. Discuss what numbers should be written along each axis to indicate the scale for each variable. Highlight the importance of the extent of the number range on each axis.

3 Have the students complete the blackline master. Ask volunteers to share their answers.

[Representations]

The Great Escape

Name _____

During daytime, a snail slithers 3 m up the inside of a well. During the night, the snail slides 1 m back down. The well is 12 m deep.

1. Complete the table below.

End of daytime	1	2	3	4	5
Height reached					

2. Use the table to help you figure out how many daytime periods the snail will need to escape the well.

   _____ daytime periods

3. Follow these steps to construct a graph to show the position of the snail after each daytime period.

   - Decide what to write on the horizontal axis (independent variable).
   - Decide what scales to use for horizontal and vertical axes.
   - Draw and label both axes.
   - Plot the points.
   - Write a title for the graph.

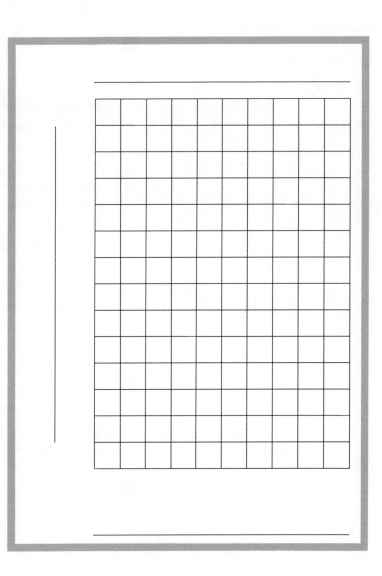

Scale It

Determining the appropriate scale on each axis of a graph

AIM

Students will determine and label the dependent and independent variable, and determine the appropriate scale for each axis of a coordinate graph.

MATERIALS

- 1 copy of the blackline master (opposite) for each student

REFLECTION

Discuss strategies for determining how to label the axes and the scale of the graphs.

1 Say, *We are buying boxes of plastic cups for the drink stand at the school fair. Cups cost $5 per box. There is a flat rate of $6 for delivery. We will need a maximum number of 12 boxes.* Draw the table shown below on the board.

Number of boxes	1	5	7	12
Total cost				

Ask, *If we buy 1 box of cups, what is the total cost?* Complete the table with the class.

2 Draw the axes of a graph on the board and then draw horizontal and vertical lines to show a 25-by-25 square grid within the axes. Ask, *Does the total cost depend on the number of boxes or does the number of boxes depend on the total cost?* Encourage the students to explain why the horizontal and vertical axes are labeled respectively as "Number of boxes" (the independent variable) and "Total cost" (the dependent variable). Ask, *What is the greatest number of boxes we can buy?* (12) *Look at the horizontal axis. Should we label the squares 1 to 10? Will our graph fill the space? What numbers should we choose for our squares?* Invite suggestions such as "label every second square from 1 to 12". Ask, *What is the greatest possible total cost?* (66) *Look at the vertical axis. How should we label the 25 squares? If we label them in 10s, will the graph fill the space? If we label them in 5s, will the graph fill the space?* The students should see that the squares along the vertical axis need to show numbers from 5 to 70 in steps of 5. Label each axis accordingly.

3 Plot the points from the table and discuss whether it makes sense to join the points. Ask, *If we buy 10 boxes, what is the total cost? If the total cost is $31, how many boxes did we buy?* Use the trend in the graph to answer the questions.

4 Together, read the story on the blackline master, and have the students complete the questions. Ask volunteers to share their answers.

[Representations]

Scale It

Name _____

An international prepaid phone card costs $5. The calls are charged at 10 cents per minute with a 50-cent connection fee.

1. Complete the table below.

Duration of call (minutes)	5	10		
Total cost of call			$2.70	$3.90

2. Use the table to help you figure out the maximum number of minutes that the card can be used.

   _____ minutes

3. Follow these steps to construct a graph for all possible phone calls on the card.
   - Decide what to write on the horizontal axis (independent variable).
   - Decide the scales for horizontal and vertical axes.
   - Draw and label both axes.
   - Plot the points.

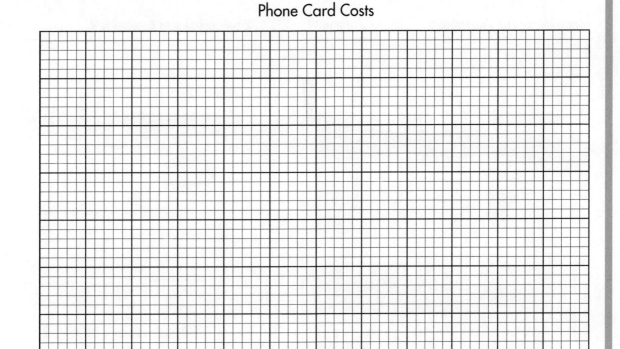

Phone Card Costs

ANSWERS

Equivalence and Equations 1 — Page 7

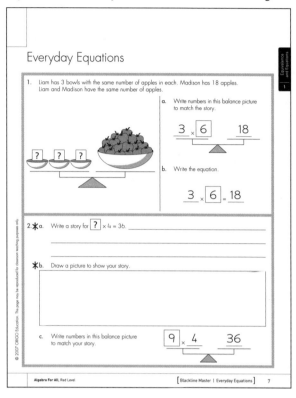

Equivalence and Equations 2 — Page 9

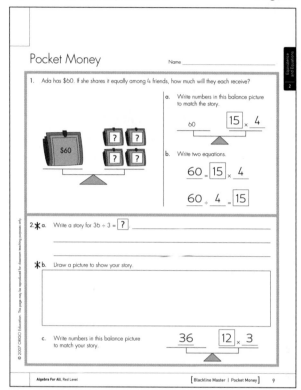

Equivalence and Equations 3 — Page 11

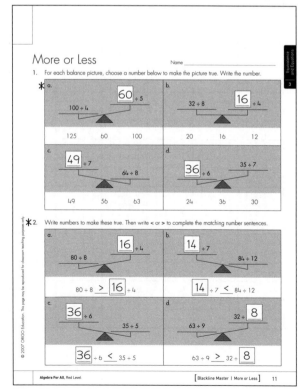

Equivalence and Equations 4 — Page 13

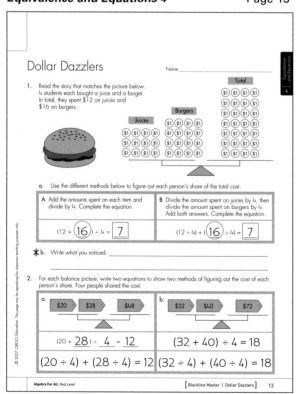

✶ Answers will vary. This is one example.

Equivalence and Equations 5 — Page 15

Keeping It Balanced

1. Complete the steps to find the number of counters in one box.

   Step 1: 20 counters; $3 \times \square + 2 = 20$
   Step 2: 18 counters; $3 \times \square = 18$
   Step 3: 6 counters; $\square = 6$

2. Follow the same steps as above to find the missing values.

   a. $3 \times \square + 4 = 25$
      $3 \times \square = 21$
      $\square = 7$

   b. $6 \times \square + 3 = 39$
      $6 \times \square = 36$
      $\square = 6$

   c. $31 = \square \times 7 + 3$
      $28 = \square \times 7$
      $4 = \square$

   d. $46 = 4 + \square \times 6$
      $42 = \square \times 6$
      $7 = \square$

   e. $5 \times \square - 6 = 29$
      $5 \times \square = 35$
      $\square = 7$

   f. $69 = \square \times 8 - 3$
      $72 = \square \times 8$
      $9 = \square$

Equivalence and Equations 6 — Page 17

Making Connections

1. A farmer wants to pump water from a tank to a house. The total length of pipe he needs is 24 m. There are 3 different lengths of pipe.

   a. Write an equation for each connection above. Then calculate the pipe lengths.

   A: $A \times 4 = 24$; Pipe A = 6 m
   B: $(6 \times 2) + (B \times 3) = 24$; Pipe B = 4 m
   C: $6 + (4 \times 3) + (C \times 2) = 24$; Pipe C = 3 m

   *b. Write how you figured it out. _____

2. The distance from a house to another tank is 32 m. Write 6 different ways to connect the tank to the house without cutting any of the pipes above. Try writing each way as an equation.

   $B \times 8 = 32$
   $(A \times 4) + (B \times 2) = 32$
   $(A \times 2) + (B \times 2) + (C \times 4) = 32$
   $A + (B \times 5) + (C \times 2) = 32$
   $(A \times 2) + (B \times 5) = 32$
   $(A \times 3) + (B \times 2) + (C \times 2) = 32$

Equivalence and Equations 7 — Page 19

Mystery Masses

For each of these, draw an object in the bag to make the picture balance. Use the clues to help you.

Equivalence and Equations 8 — Page 21

Boxed In

A store sells different-shaped boxes. Write prices in the shapes to make the equations true.

1. $\square + \bigcirc = \$47$; $\square - \bigcirc = \$7$

   Write or draw how you figured out the cost of these boxes.
   $\square + \bigcirc + \square - \bigcirc = \$47 + \$7$
   $\square + \square = \$54$
   $\square = \$27$
   $\$27 - \bigcirc = \$7$
   $\bigcirc = \$20$

2. $\square + 2 \times \hexagon = 15$; $\square + 5 \times \hexagon = 33$

   Write or draw how you figured out the cost of these boxes.

$\square$	1	3	5	7	9
$\hexagon$	7	6	5	4	3
$\square + 2 \times \hexagon = 15$	15	15	15	15	15
$\square + 5 \times \hexagon = 33$	36	33	30	27	24

   Therefore, $\square = 3$, $\hexagon = 6$

* Answers will vary. This is one example.

ANSWERS

Patterns and Functions 1 Page 23

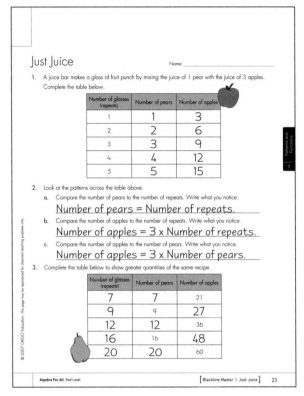

Patterns and Functions 2 Page 25

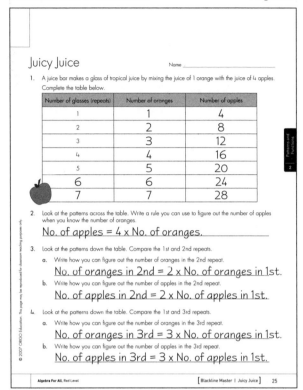

Patterns and Functions 3 Page 27

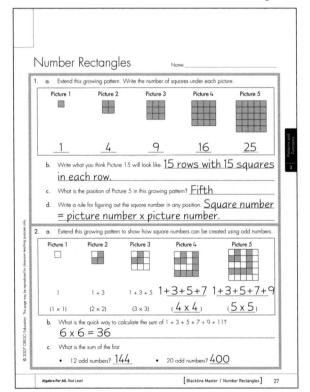

Patterns and Functions 4 Page 29

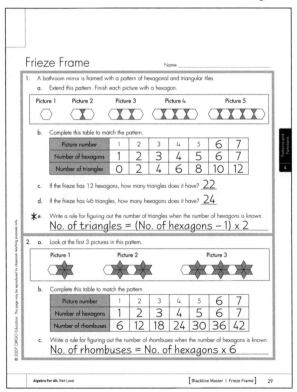

* Answers will vary. This is one example.

Patterns and Functions 5 Page 31

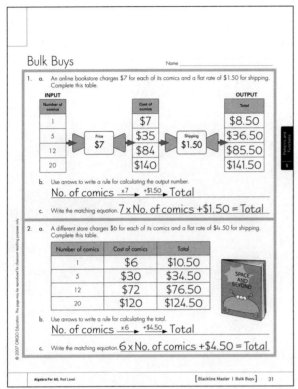

Patterns and Functions 6 Page 33

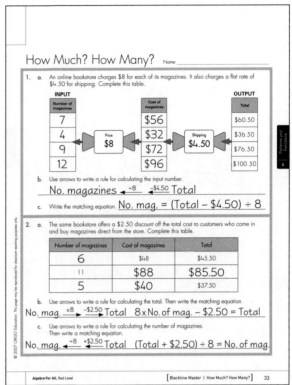

Patterns and Functions 7 Page 35

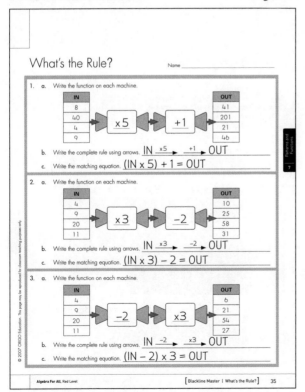

Patterns and Functions 8 Page 37

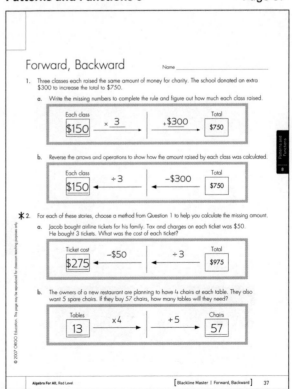

* Answers will vary. This is one example.

ANSWERS

Patterns and Functions 9 — Page 39

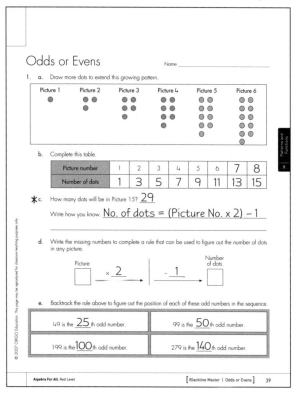

Properties 1 — Page 41

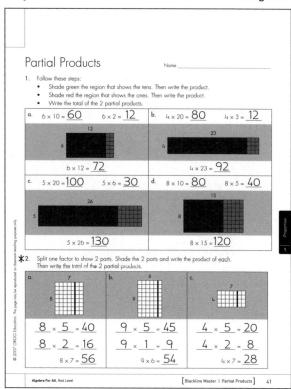

Properties 2 — Page 43

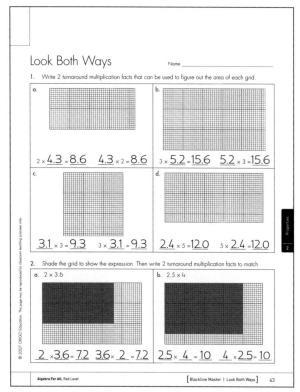

Properties 3 — Page 45

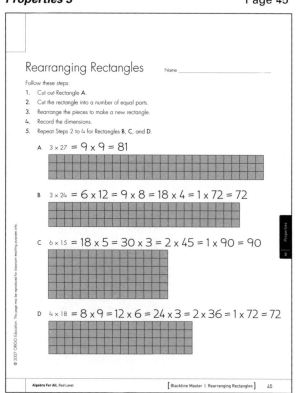

* Answers will vary. This is one example.

Properties 4 — Page 47

No Change

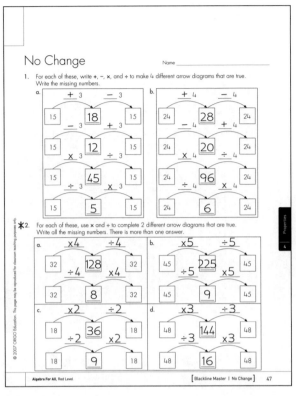

Properties 5 — Page 49

Following Orders

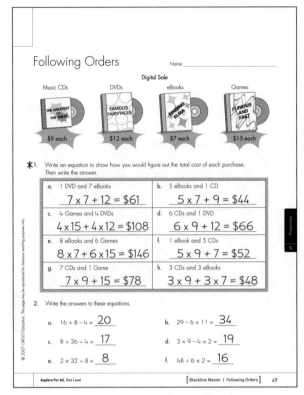

Properties 6 — Page 51

Building Blocks

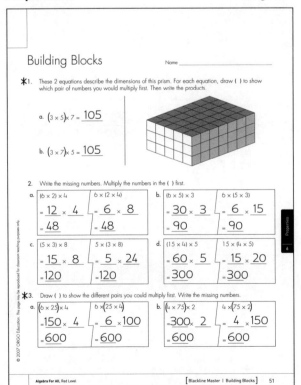

Representations 1 — Page 53

High in the Sky

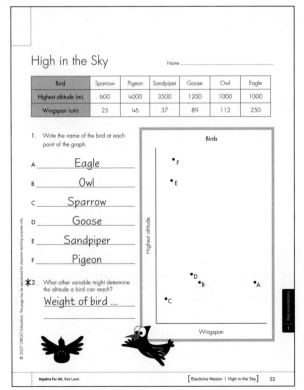

* Answers will vary. This is one example.

Algebra For All, Red Level

ANSWERS

Representations 2 — Page 55

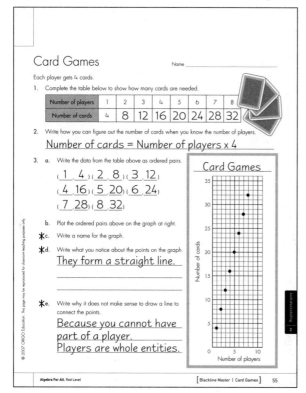

Representations 3 — Page 57

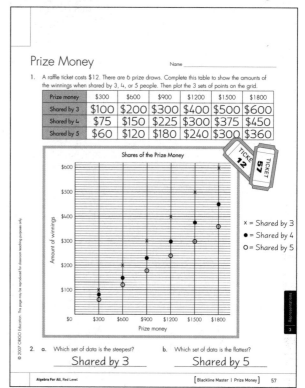

Representations 4 — Page 59

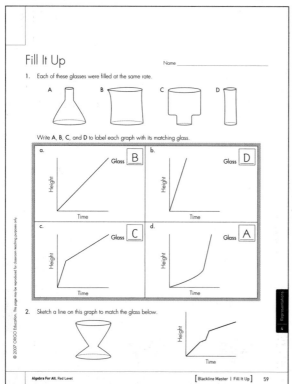

Representations 5 — Page 61

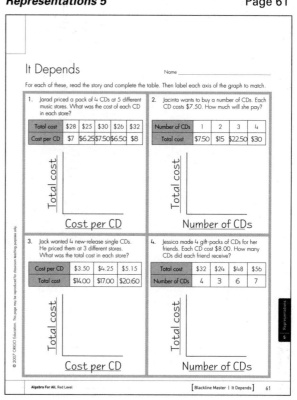

* Answers will vary. This is one example.

Representations 6 — Page 63

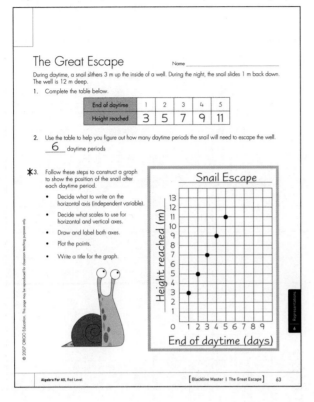

Representations 7 — Page 65

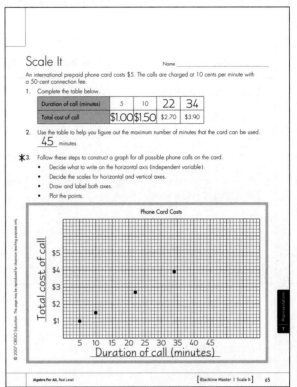

* Answers will vary. This is one example.

Assessment Summary

Name _____

	Lesson	Page	A	B	C	D	Date
Equivalence and Equations	Everyday Equations	6					
	Pocket Money	8					
	More or Less	10					
	Dollar Dazzlers	12					
	Keeping it Balanced	14					
	Making Connections	16					
	Mystery Masses	18					
	Boxed In	20					
Patterns and Functions	Just Juice	22					
	Juicy Juice	24					
	Number Rectangles	26					
	Frieze Frame	28					
	Bulk Buys	30					
	How Much? How Many?	32					
	What's the Rule?	34					
	Forward, Backward	36					
	Odds or Evens	38					
Properties	Partial Products	40					
	Look Both Ways	42					
	Rearranging Rectangles	44					
	No Change	46					
	Following Orders	48					
	Building Blocks	50					
Representations	High in the Sky	52					
	Card Games	54					
	Prize Money	56					
	Fill It Up	58					
	It Depends	60					
	The Great Escape	62					
	Scale It	64					

Algebra for All, Red Level